AF601244

Current Trends in Farming Systems Research

The Authors

Dr. R.K. Nanwal is working as Chief Scientist in Department of Agronomy, CCSHAU, Hisar. He got ICAR team research award for the triennium 1994-96 & CCSHAU best teacher award in 2005-06. He has written 7 books, 12 manuals and more than 150 research papers in journals of repute.

Dr. A.K. Mehta is working as Principal Scientist in Farming System Project. He has a long experience in extension as well as in farming systems as OFR agronomist.

Dr. Manoj Kumar is working as Principal Soil Scientist in Department of Agronomy. He has more than 10 year experience in Integrated Farming Systems.

Dr. V.S. Hooda is working as Assistant Agronomist under AICRP on IFS. His major contribution is in the field of weed control, farming systems and teaching.

Dr. Naval Kishor Kamboj is working as Senior Research Fellow under AICRP on IFS in Department of Agronomy. He did his master's degree from YSPUHF, Solan (H.P.) and Ph.D. from CCSHAU, Hisar.

Dr. Jagdev Singh is working as Head Department of Agronomy in CCSHAU, Hisar. He has published 5 bulletins and more than 90 research papers in journals of repute.

Current Trends in Farming Systems Research

Compiled by

R.K. Nanwal

A.K. Mehta

Manoj Kumar

V.S. Hooda

Naval Kishor Kamboj

Jagdev Singh

Department of Agronomy
CCS Haryana Agricultural University
Hisar-125004 (Haryana)

2017

Scholars World

A Division of

Astral International Pvt. Ltd.

New Delhi – 110 002

Cataloging in Publication Data--DK
Courtesy: D.K. Agencies (P) Ltd. <docinfo@dkagencies.com>

Current trends in farming systems research / compiled by R.K. Nanwal [and five others].
pages cm
Includes bibliographical references.
ISBN 9789387057418 (International Edition)

1. Agricultural systems--Research--India. I. Nanwal, R. K., editor.

S494.5.S95C87 2017 DDC 630.72054 23

Published by : **Scholars World**
A Division of
Astral International Pvt. Ltd.
– ISO 9001:2015 Certified Company –
4736/23, Ansari Road, Darya Ganj
New Delhi-110 002
Ph. 011-4354 9197, 2327 8134
E-mail: info@astralint.com
Website: www.astralint.com

Acknowledgement

Farming system is the way of life for small and marginal holders. Recently the AICRP on Farming System Research at CCSHAU, Hisar, has developed various technologies on regional based cropping systems, integrated nutrient management in pearl millet-wheat cropping system, organic farming package in mungbean-wheat (*desi*) cropping system, resource conservation technologies *etc.* Integrated faming system model developed in one hectare area gives hope for realizing higher returns for farmers from a unit land. Family farming model of one hectare developed at on station Hisar gives decent round the year income with more than 2 lakh as return over variable cost/annum. The on farm research in Sirsa district also intensified and regular scientist-farmer interaction, training and demonstrations were also organized. In the coming years, on farm research in farmer participatory IFS approach and capacity building of stake holders are to be given thrust.

The reported information based on synthesis of long term data clearly reveals that the yield of crops and system can be enhanced substantially through various agronomic manipulations. The projective indices could help in resource optimization for improving the livelihood security of the farmers. The research bulletin titled **"Current trends in Farming System Research"** will be quiet helpful for policy makers, researchers, students and farming community to address the issues of farming system. The bulletin has been written in a lucid language for easy understanding of the concept.

We express our indebtedness to Dr K.P. Singh, Worthy Vice-Chancellor, CCSHAU, Hisar for his keen interest in the farming system research programme. The task of research bulletin writing could not have been possible but due to encouragement by him, it was fulfilled. We thankfully acknowledge Dr. A.S. Panwar, Director, IIFSR, Modipurum for his moral support and financial help rendered to the project. We are gratefully to Dr. N. Ravisankar, Programme Facilitator, IIFSR, Modipurum for providing full assistance to the project in time. Our thanks are due particularly to Dr. Jagdev Singh, Professor and Head, Department of Agronomy, who is also one of the authors for the guidance during the course of the writing of the bulletin. The authors feel obliged to the scientist namely Drs. S.K. Yadav, Hari Om, Pawan Kumar, Mahesh Kumar and Shewta for their contribution in conducting various experiments in the project during the period under report.

Authors

Vice-Chancellor
CCSHAU, Hisar

K.P. Singh

Foreword

The glory of Green Revolution proved to be a turning point in improving the food situation from "begging bowl to self-sufficiency". Now days several emerging issues have been addressed to sustain the high productivity under different cropping and farming systems. The future of Indian agriculture depends on the development of appropriate farming system as applicable to resource poor farm families and as suited to small and marginal farmers of different agro-ecological zones. Integrated farming system is not only a reliable way of obtaining fairly high productivity with substantial fertilizer economy, but also a concept of ecological soundness leading to sustainable agriculture.

A judicious mix of cropping systems with associated enterprises like dairy, poultry, piggery, fishery, sericulture, bee keeping, mushroom production, vermicompost unit etc suited to the given agro- climatic conditions and socio-economic status of the farmers would bring prosperity to the farmer. The goal of sustainable agriculture is to maintain production at levels necessary to meet the increasing demand of an expanding world population without degrading the environment. The concept of sustainability applied to agriculture developed mainly as a result of growing awareness of negative impacts of intensive farming systems on the environment and the quality of life of rural and neighboring communities. The book is comprehensive covering current trends in farming systems research conducted under AICRP on FSR at CCSHAU, Hisar.

The authors of the book did a commendable job by compiling and presenting the findings of the experiments conducted both at on station and on farm in a systematic manner so as to be useful to the readers. It is praise worthy effort and it is hoped that the purpose for which this compilation has been published will be fully served and will stimulate further research work in this important area of agronomy.

K.P. Singh

Contents

Chapter 1

Introduction

The glory of Green Revolution proved to be a turning point in improving the food situation from "begging bowl to self-sufficiency". The future of Indian agriculture depends on the development of appropriate farming system as applicable to resource poor farm families and as suited to different agro-ecological zones. A judicious mix of cropping systems with associated enterprises like dairy, poultry, piggery, fishery, sericulture, bee keeping, mushroom production etc suited to the given agro- climatic conditions and socio-economic status of the farmers would bring prosperity to the farmer. Integrated farming system is not only a reliable way of obtaining fairly high productivity with substantial fertilizer economy, but also a concept of ecological soundness leading to sustainable agriculture. Indian Agriculture has made very remarkable strides in the last four decades. Introduction of input responsive, high yielding, photo-insensitive and short duration varieties coupled with improved production technology ushered the era of high crop productivity.

The rapidly rising food needs due to steadily mounting demographic pressures necessitated systems approach in an integrated manner. It is how multiple/intensive cropping came into existence. Commensurating identification of efficient crops, evaluation of crop varieties and formulation of agro-practices landed adequate support to our war on low productivity and India was able to witness Green Revolution to belie the prophets of dooming. In fact, the Indian Agriculture, therefore, never looked back. Traditional cropping systems were re-tailored for intensification both in time and space.

Farming System

The process of harnessing solar energy in the form of economic plant and animal product is called farming. A set of inter-related practices/processes organised into a functional entity is called a system. Farming system is a set of agricultural activities organised while preserving land productivity, environmental quality and maintaining desirable level of biological diversity and ecological stability. Farming system is a complex inter-related matrix of soil, plants, animals, implements, power, labour, capital and other inputs controlled in part by farm families and influenced by varying degrees of political, economic, institutional and social forces that operate at many levels. It is a set of elements or components that are interrelated which interact among themselves. Farmer exercises control and choice regarding the type and result of interaction.

It represents integration of farm enterprises such as cropping systems, animal husbandry, fisheries, forestry, sericulture, poultry etc for optimal utilization of resources bringing prosperity to the farmer. The farm products other than the economic products, for which the crops are grown, can be better utilized for productive purposes in the farming systems approach.

Farming Systems Concept

In farming system, the farm is viewed in a holistic manner. The farmers are subjected to many socio-economic, bio-physical, institutional, administrative and technological constraints. A combination of one or more enterprises with cropping when carefully chosen planned and executed, gives greater dividends than a single enterprise, especially for small and marginal farmers.

Farm as a unit is to be considered and planned for effective integration of the enterprises to be combined with crop production activity, such that the end-products and wastes of one enterprise are utilized effectively as inputs in other enterprise. For example the wastes of dairy component *viz.*, dung, urine, refuse etc are used in preparation of FYM or compost which serves as an input in cropping system. Likewise the straw obtained from crops (maize, rice, sorghum etc) is used as a fodder for dairy cattle. Further, in sericulture the leaves of mulberry crop are used as a feeding material for silkworms, grains from maize crop are used as a feed in poultry etc. The selection of enterprises must be based on the cardinal principle of minimizing the competition and maximizing the complementarity between the enterprises.

Sustainability is the objective of the farming system where production process is optimized through efficient utilization of inputs without compromising on the quality of environment with which it interacts on one hand and attempt to meet the national goals on the other. The concept has an undefined time dimension. The magnitude of time dimension depends upon ones objectives, being shorter for economic gains and longer for concerns pertaining to environment, soil productivity and land degradation.

Principles of Farming System

- Minimization of risk
- Recycling of wastes and residues
- Integration of two or more enterprises
- Optimum utilization of all resources
- Maximum productivity and profitability

- ☆ Ecological balance
- ☆ Generation of employment potential
- ☆ Increased input use efficiency
- ☆ Use of end products from one enterprise as input in other enterprise

Characteristics of Farming System

1. Farmer oriented and holistic approach
2. Effective farmer's participation
3. Unique problem solving system
4. Maintains the long-term biological and ecological integrity of natural resources
5. Sustain a desirable level of support to a farm's, communities or regions social, political, and economic well being
6. Dynamic system
7. Gender sensitive
8. Responsible to society
9. Environmental sustainability
10. Location specificity of technology
11. Diversified farming enterprises to avoid risks due to environmental constraints
12. Provides feedback from farmers
13. Enhances quality of life.

Goals of Integrated Farming System

The four primary goals of IFS are:

- ☆ Maximization of yield of all component enterprises to provide steady and stable income at higher levels
- ☆ Rejuvenation/amelioration of system's productivity and achieve agro-ecological equilibrium
- ☆ Control the buildup of insect-pests, diseases and weed population through natural cropping system management and keep them at low level of intensity
- ☆ Reducing the use of chemical fertilizers and other harmful agro-chemicals and pesticides to provide pollution free, healthy produce and environment to the society at large

Advantages of Farming System

1. **Productivity-** In farming system, there is an opportunity to increase economic yield per unit area per unit time by virtue of intensification of crop and allied enterprises. Time concept by crop intensification and space

concept by building up of vertical dimension through crops and allied enterprises are the means of enhancing productivity.

2. **Profitability** – There is also an opportunity to make use of produce/waste material of one enterprise as an input in another enterprise at low/no cost. Thus with reduction in production cost, the profitability and benefit cost ratio works out to be high.

3. **Potentiality** - Soil health is getting deteriorated and polluted due to faulty agricultural management practices *viz.*, imbalanced use of inorganic fertilizers, dumping of pesticides, poor irrigation management etc. In farming system, there is organic supplementation through effective use of manures and waste recycling, thus providing an opportunity to sustain production for a longer time.

4. **Balanced food-** In farming system, diverse enterprises are involved for production of different sources of nutrition *i.e.* proteins, carbohydrates, fats and minerals etc form the same unit area. It is helpful in solving the malnutrition problem of vegetarian families of marginal and sub-marginal farmers.

5. **Environmental safety-** Farming system makes use or conserves the by product/waste product of one component as input in another component and use of bio-control measures for pest and disease control. These eco-friendly practices reduce the application of huge quantities of fertilizers and pesticides lowering the pollution of soil water and environment.

6. **Cash flow round the year-** Unlike conventional single enterprise crop activity where the income is expected only at the time of sale of produce after harvesting of the crop, there is cash flow round the year by way of sale of products from different enterprises of IFS *viz.*, eggs from poultry, milk from dairy, fish from fisheries, silkworm cocoons from sericulture, honey from apiculture etc. Purchasing power of the farmer is improved and he can invest in improved technologies for enhanced production.

7. **Energy saving-** Fossil fuel is not going to stay forever. Rather it is declining at a faster rate leading to a situation where in the whole world may suffer for want of fossil fuel. In farming system, effective recycling of organic wastes is helpful in generating energy from biogas plants to mitigate energy crisis to some extent.

8. **Meeting fodder crises-** In IFS there is effective utilization of whole of land area. Growing perennial fodder legume along the border or water courses, intensification of cropping including fodder legumes in cropping systems helps to produce the required fodder and greatly relieve the problem of non availability of fodder to livestock even during lean period.

9. **Solving timber and fuel crises-** Current production level of 20 million m3 of fuel wood and 11 million m3 of timber wood is no match for the demand estimated of 360 m3 of fuel and 64.4 million m3 of timber wood. Hence, the current production needs to be stepped up several-folds.

Afforestation programmes in addition to introduction of agro-forestry component in farming system without detrimental effect on crop yield will greatly reduce deforestation and preserve our natural ecosystem.

10 **Employment generation-** Various farm enterprises *viz.*, crop +livestock or any other allied enterprise in the farming system would increase labour requirement significantly and would help solve the problem of under employment. IFS provides enough scope for providing employment to family labour round the year.

11. **Scope for establishment of agro-industries-** Once the produce from different components in IFS is increased to a commercial level there will be surplus produce for value addition in the region which may lead to establishment of agro-industries.

12. **Enhancement in input use efficiency –** An IFS provides good scope for resource utilization in different components leading to greater input use efficiency and benefit- cost ratio.

13. **Improvement in living standard of farmer-** IFS is helpful in improvement of living standard by way of increased income round the year from different enterprises. Farmers will also have a feeling of security. They will be able to allocate resources for attaining their family goal.

Role of Farming Systems in Sustainable Agriculture

In India progress in agriculture was through exploitation of recourses to serve the interest of rules in the power. In the post- independence era, we were faced with the difficult task of feeding an increasing population and were burned by recurring import of food grains. Attending self- sufficiency in food grain production was the challenges and a goal before agricultural scientists. The element of strategy to achieve self- sufficiency in food production within a short time were to expand the area under cropping, make large investments in development of water resources, develop input supplies and market infrastructure. Developments and promotion of high yielding varieties responsive to high inputs was the cornerstone for green revolution.

We are self sufficient in food grain production. Now the question arises whether to follow the same strategies, which resulted in green revolutions or to redefine them. Consideration of our past achievement new challenges and the related development worldwide can answer this question.

1. Although we are now self sufficient in food grain production, the per capital availability has not increased significantly and a large fraction of population are still below poverty line due to lack of purchasing capacity.
2. At the present growth rate of 2 per cent per annum, we will be about 1.3 billion by 2025 and 1.6 billion by 2060, hence there is need to produce more from shrinking resources base.
3. Major gains in productivity and production in the past three decades have been from areas, which had no serious limitation to production (adequate

irrigation productive soils, ideal climate etc.) productivity and production increases was moderate in eastern and central India and Deccan plateau.

4. Even in areas with no serious limitation for production, technology adopted for enhancing productivity has also simultaneous weaken the resource base resulting in a series of second – generation problems.

Modern crop production technology has considerably raised output but has created problems of land degradation, pesticide residue in farm produce, erosion, and atmosphere and water pollution. The natural resources base is degraded and diminished and the quality of the environment sustaining human life is adversely affected. Agricultural production has sustained man and great civilization. The history of the world reveals that great civilization flourished along irrigation sources and mismanagement of these resources saw the extinction of these civilizations. With expanding population and rapid depletion and degradation of the natural resource base sustainable agriculture has assumed very great significance. The task of meeting the needs of the present generation without eroding the ecological assets of the future generation is receiving up priority by environmental planners.

Recent Advances in Farming Systems in India

The Green Revolution in India has achieved self-sufficiency in food production. However, in the state of Haryana this has resulted in continuous environmental degradation, particularly of soil, vegetation and water resources. Soil organic matter levels are declining and the use of chemical inputs is intensifying. Newly introduced crop varieties have been responsive to inputs but this has necessitated both increased fertiliser application and use of irrigation resulting in water contamination by nitrate and phosphate and changes in the ground water table. With 82 per cent of the geographic area already under cultivation, the scope for increased productivity lies in further intensification which is crucially dependent on more energy-intensive inputs. Declining nutrient-use efficiency, physical and chemical degradation of soil, and inefficient water use have been limiting crop productivity, whilst the use of monocultures, mechanisation and an excessive reliance on chemical plant protection have reduced crop, plant and animal diversity in recent years. About 60 per cent of the geographical area faces soil degradation (waterlogging, salinity and alkalinity) which threatens the region's food security in the future. Since 1985, the water table has risen more than 1 m annually, and patches of salinity have started to appear at the farm level. The situation is worse in higher rainfall areas where waterlogging follows shortly after the rains. Apart from affecting agricultural crops, a high water table causes floods even following slight rains due to the reduced storage capacity of the soil. Such ecological impacts are motivating farmers to reduce fertiliser and pesticides use. This has led to an increased investment in alternative technology and products including an interest in Integrated Pest Management.

There has been a remarkable shift in India in the cropping patterns for both rainy season and winter season since the Green Revolution. Rice (Oryza sativa) and wheat (Triticum aestivum) have replaced pulses, bajra (Pearl millet), jowar/ sorghum (Syricum), as dominant food crops, while cotton (Gossypium spp.) is the key cash crop. The main rainy season crops in 1965–1966 comprised bajra (46 per cent), rice (13 per cent) and sorghum (12 per cent); however in 1995–1996 rice (34 per cent) was the major crop followed by bajra (27 per cent) and cotton (24 per cent). For winter crops wheat has increased in production as major crop from 43 per cent in 1965–1966 to 64 per cent in 1995–1996. In Haryana, the yields of rice and wheat have increased considerably. Cropping patterns have changed as a result of the new agricultural technologies including irrigation facilities, improved seed varieties, pesticides, insecticides and new methods of farming. Recently, however, the two dominant crops, rice and wheat, have been facing severe constraints and challenges for sustainable productivity.

New Issues

- ☆ Faced with the critical situation of intensive monocultures, new conceptual ways of constructing sustainable agroecosystems are being sought. Several agronomists recently proposed that traditional multispecies systems could be used as models for designing sustainable cropping systems. Role of woody perennial species enhanced the sustainability of ecosystem functioning in the humid tropics and proposed forest-like agro ecosystems. Such systems are usually complex, as they are based on several species, and may involve combinations of perennial and annual, woody and nonwoody plants.
- ☆ Agricultural research now has an adequate tool-box of methods and models for technology development in monospecific cropping systems, but its suitability for more complex systems is unsure. Methods for designing multispecies systems barely exist. Systemic agronomy concepts (crop management sequences, cropping system), and especially the tools derived from that discipline, scarcely deal with the complexity of multispecies systems. In particular, the modeling tools widely used today in agronomy are not well adapted to simulating them. New models are required to represent, assess and design sustainable multispecies cropping systems.

Research efforts in the past have been directed not only on increasing productivity of individual crop, but also towards improving the efficiency of farm utilization or maximizing return per unit area per unit time. The results of these experiments have come up with suitable cropping patterns for different areas with annual food crops, as the base around which the cropping patterns are built.

Chapter 2

Brief Introduction of AICRP on Farming Systems Research

Work of the All India Coordinated Research Project (AICRP) was started in 1953 as Soil Fertility Research Programme. As a full-fledged coordinated project, the programme was launched in 1968. In earlier years, the experiments on cultivator's fields were mostly restricted to fertilizer and varietal trials. At the research stations, experiments were planned to study the production potential of cropping systems under adequate and limited resources and to make judicious use of inputs like fertilizers, pesticides, irrigation, seeds etc. The main objective of the project was to develop, continuously update and test on cultivator's fields, the technology for various crop based farming systems.

This project (AICRP) was initiated with two components namely Model Agronomic Experiments and Experiments on Cultivator's Fields (ECF). In VIIth Five Year Plan in 1983, the Model Agronomic Centre was re-named as Cropping Systems Research Centers to cover on station research and on farm research. The programme has been remodeled in 2008-09 and has been renamed as AICRP on Integrated Farming Systems.

The AICRP on "Integrated Farming Systems" is being in operation at CCS Haryana Agricultural University, Hisar since its inception. It is financed by 75 per cent share from ICAR and 25 per cent from State Government. The project has two major components *viz.* (i) On Station Research (ii) On Farm Research *i.e.* experiments on cultivator's field.

Haryana State is situated at latitude 27 39′ to 30 55′ and longitude 73 8′ to 77 36′ E, comprises a part of the Indo-Gangetic Plain. It has common borders on south and west with Rajasthan, on the north with Punjab and Himachal Pradesh and on the east Uttar Pradesh and Delhi.

In Haryana, the main centre of farming systems research project is situated at Hisar. Hisar is situated in the south-western zone of the State. It has arid tropical climate which is influenced by westerly winds in summer touching the temperature as high as 49 C, whereas in winter, north-westerly winds bring down the temperature even below 0 C. Annual rainfall in the district varies between 300–450 mm with coefficient of variation of about 45 per cent. Annual potential evapotranspiration ranges between 1600 and 1650 mm. The soils are loamy sand to sandy loam in texture and water table is high.

The on farm experiments (OFR) centre of the project is situated at Sirsa in south-western zone of the State and it also experiences extremes of temperature in summer and winter. In summer, the maximum temperature goes beyond 49 C while in winter it goes below 0 C. Seventy five per cent of the total rainfall is received during the months from July to September and winter rains are received during December to February. The soil is sandy loam in texture.

The long-term objective of the farming systems research project is to develop intensive and sustainable farming systems with increased productivity and efficient resource use in different agro-climatic regions of the State. The specific objectives of the project are:

1. To develop profitable and energy efficient cropping systems for different agro-climatic regions with special emphasis on oilseeds, pulses and fodder crops.
2. To develop integrated nutrient supply and management systems for major cropping systems with emphasis on locally available resources.
3. To improve fertilizers use efficiency in cropping systems through the use of efficient carriers, amendments and residual effects.
4. To get maximum yield and remuneration with the use of organic sources.
5. To study the effect of different tillage and planting techniques in different cropping systems to improve crop productivity and soil properties.
6. To develop integrated farming system model that generate maximum returns to the farmers.

The mandate of farming systems research project is as under:

1. Identification of need based cropping system for different agro-climatic conditions.
2. Permanent experiment on integrated nutrient supply system on cereal based crop sequence.
3. Development of organic farming package for mungbean-wheat (*desi*) cropping system.

4. Resource conservation technology in pearl millet–mustard/wheat cropping systems.
5. To identify the most remunerative integrated farming system model.

The mandate for on Farm Research is as under:

1. Nutrient management in cotton-wheat cropping system
2. Diversification of existing farming system
3. On farm evaluation of farming system modules for improving the profitability of small and marginal farmers

Chapter 3

Weather Report

Rainfall

The year 2010, 2013 and 2015 were good rainfall years in the decade (2005-06 to 2015-16). A total rainfall of 810.1 mm was received during 2013 (Table 1) which was highest among all the years. It was followed by 774 and 641.4 mm during the year 2010 and 2015, respectively. Average rainfall around 500 mm was received during the year 2007, 2008, 2009, 2011and 2012, which was close to the average rainfall of Hisar. The rainfall was deficient during the year 2006 (346.1 mm).

In general, rainfall was higher during *Kharif* followed by *Rabi* and summer season. The rainfall during *Kharif*, *Rabi* and summer was 391.7, 158.6 and 91.1 mm, respectively during the year 2015

Temperature

The climate of Hisar is semi-arid and subtropical with hot and dry winds during summer months. Warm humid in monsoon and cold dry weather in winter are the general features of this region. Both, winter and summer are usually harsh to bear upon. The mean minimum and maximum temperature exhibit wide range. A maximum temperature zooming 44 to 48°C during summer and temperature dipping as low to freezing point accompanied with chill frost in winter is of common occurrence. The year 2010 was a very hot year as maximum temperature was recorded highest during May month over rest of the years. The year 2007 and 2008 were considered to be cold years as minimum temperature during these two years ranged between 3.2-3.3°C.

Relative Humidity

Mean relative humidity attains the maximum value 85 to 90 per cent during the south-west monsoon and the minimum 30 to 45 per cent during the summer

Table 1: Meteorological Data of Rainfall (mm) for the Period of 2006-2015 at Hisar

Month	*Rainfall (mm)*									
	2006	*2007*	*2008*	*2009*	*2010*	*2011*	*2012*	*2013*	*2014*	*2015*
January	0.0	0.0	3.3	10.3	11.5	0.0	14.4	43.0	2.0	15.4
February	0.0	86.1	0.9	6.1	7.6	34.8	0.0	32.7	12.5	12.2
March	27.2	44.3	0.0	4.1	2.5	12.5	0.0	31.1	47.0	121.1
April	0.0	2.0	13.9	24.9	0.0	35.2	33.3	2.3	16.4	91.1
May	69.2	46.2	52.7	38.2	1.9	84.9	29.8	0.0	56.5	0.0
June	74.7	167.3	108.9	29.4	50.3	57.0	26.5	97.3	71.6	161.0
July	91.3	21.0	162.1	92.4	300.0	82.5	76.6	159.2	73.1	156.1
August	7.9	68.8	129.1	14.0	209.9	95.7	282.5	288.2	34.2	54.8
September	69.8	66.3	96.0	239.9	147.6	141.1	32.9	140.4	81.5	19.8
October	0.0	0.0	5.3	0.0	0.0	0.0	5.4	6.5	21.3	7.0
November	0.0	0.0	3.2	0.0	0.0	0.0	0.0	9.4	0.0	2.9
December	6.0	3.8	0.8	0.0	43.6	0.0	5.5	0.0	9.0	0.0
Total	346.1	505.8	576.2	459.3	774.9	543.7	506.9	810.1	425.1	641.4

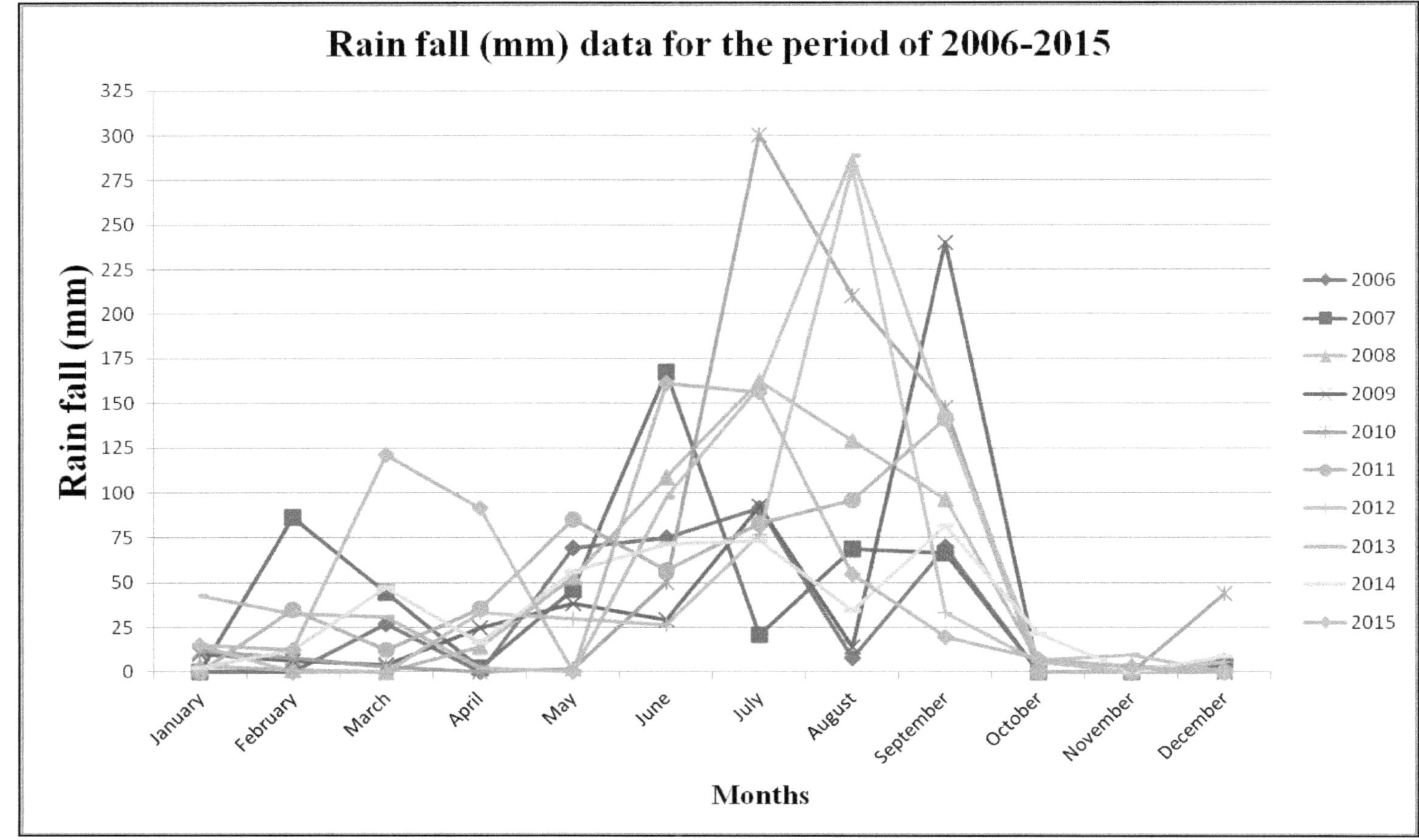
Rain fall (mm) data for the period of 2006-2015
Rain fall (mm)
325
300
275
250
225
200
175
150
125
100
75
50
25
0
January
February
March
April
May
June
July
August
September
October
November
December
Months
2006
2007
2008
2009
2010
2011
2012
2013
2014
2015

Table 2: Mean Meteorological Data of Temperature (°C) for the Period of 2006-2015 at Hisar

Month	*Temperature (°C)*																			
	2006		*2007*		*2008*		*2009*		*2010*		*2011*		*2012*		*2013*		*2014*		*2015*	
	Max.	*Min.*	*Max.*	*Min.*	*Max.*	*Min.*	*Max.*	*Min.*	*Max.*	*Min.*	*Max.*	*Min.*	*Max.*	*Min.*	*Max.*	*Min.*	*Max.*	*Min.*	*Max.*	*Min.*
January	20.8	4.0	20.1	3.3	18.6	3.2	20.1	5.8	17.3	5.9	16.9	4.2	18.4	4.8	17.6	4.2	18.0	5.6	16.2	6.0
February	27.4	9.9	22.8	8.8	21.2	4.2	24.1	7.3	25.8	7.4	22.7	8.1	21.0	5.3	21.5	8.9	20.8	7.6	23.5	9.6
March	27.8	11.4	27.0	11.2	32.3	12.0	29.4	11.8	34.9	16.8	28.6	11.4	28.5	10.3	28.4	12.0	26.3	12.2	26.5	12.4
April	37.0	18.0	38.1	17.7	35.4	17.1	35.8	17.3	41.1	20.4	34.4	16.7	34.1	18.0	35.0	17.2	34.1	17.1	34.0	19.2
May	40.6	23.9	39.8	23.8	38.0	22.7	40.8	23.7	42.9	24.4	40.1	23.9	39.6	21.9	41.5	22.7	38.4	22.2	40.4	23.0
June	38.1	24.2	38.5	26.5	36.2	25.6	41.5	25.2	40.5	25.8	38.9	26.2	41.8	27.9	39.3	27.2	41.0	26.3	38.2	25.0
July	35.4	26.8	36.9	26.4	36.2	26.4	36.7	26.4	35.3	26.4	35.6	26.3	38.3	27.9	36.6	26.9	37.4	27.5	34.5	26.0
August	35.0	25.5	35.3	26.2	33.7	25.4	38.0	26.8	33.8	26.0	34.1	25.8	33.6	26.1	33.3	25.9	36.4	26.3	34.7	26.1
September	34.2	22.7	34.0	23.7	33.6	22.4	34.4	22.5	32.2	23.3	33.3	23.1	33.5	24.0	34.4	23.9	34.7	23.8	35.8	22.6
October	33.7	17.3	33.7	13.9	33.9	18.8	33.4	15.5	33.2	18.1	33.0	15.4	32.6	15.3t	31.9	20.1	33.2	18.5	34.3	18.5
November	28.5	10.9	29.4	10.6	28.9	10.6	27.1	9.9	27.7	11.5	29.4	11.0	27.6	9.3	26.6	10.4	28.2	10.2	27.8	12.3
December	21.9	5.5	21.5	4.8	23.5	6.8	23.1	4.9	21.3	4.6	22.9	5.2	21.2	6.1	21.8	7.0	19.5	6.0	22.4	6.0

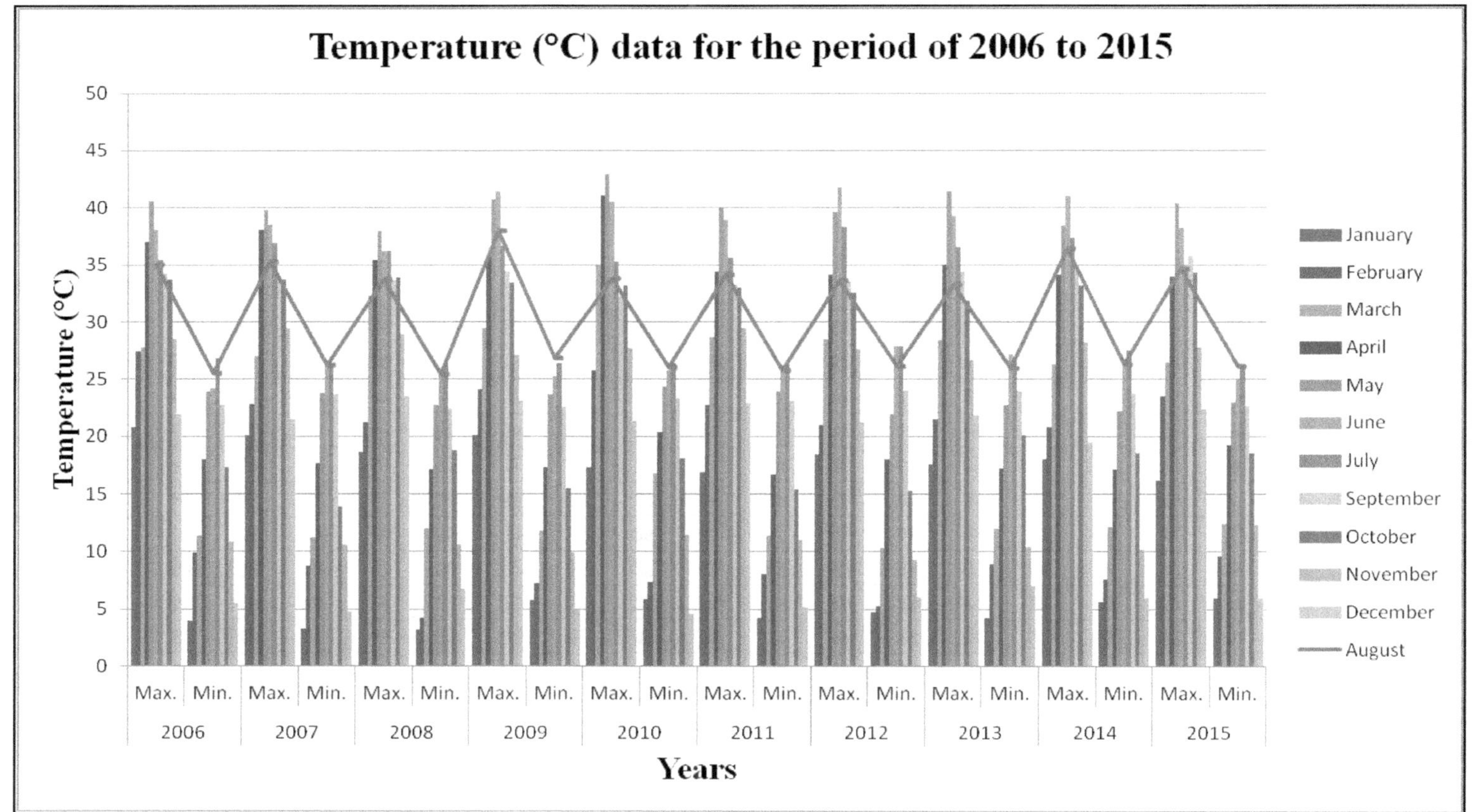
Temperature (°C) data for the period of 2006 to 2015
Temperature (°C)
50
45
40
35
30
25
20
15
10
5
0
Max.
Min.
2006
2007
2008
2009
2010
2011
2012
2013
2014
2015
Years
January
February
March
April
May
June
July
September
October
November
December
August

Table 3: Meteorological Data of Relative Humidity (Per cent) for the Period of 2006-2015 at Hisar

Month	*Relative Humidity (Per cent)*																			
	2006		*2007*		*2008*		*2009*		*2010*		*2011*		*2012*		*2013*		*2014*		*2015*	
	M	*E*	*M*	*E*	*M*	*E*	*M*	*E*	*M*	*E*	*M*	*E*	*M*	*E*	*M*	*E*	*M*	*E*	*M*	*E*
January	92	42	94	43	90	38	97	53	98	69	94	53	96	51	95	58	97	69	97	73
February	94	50	96	72	86	37	93	45	91	45	95	52	87	41	96	60	95	67	91	54
March	92	49	90	45	86	30	89	40	78	31	92	45	84	32	92	47	90	55	92	54
April	66	29	66	23	63	26	62	30	51	18	80	43	74	38	68	27	72	35	72	35
May	68	36	59	27	60	32	55	26	47	18	59	30	52	24	48	17	66	30	58	26
June	72	45	72	46	80	59	55	25	60	33	69	41	53	27	69	44	61	34	73	49
July	85	62	81	52	82	58	78	58	88	66	85	61	76	51	82	59	77	51	86	64
August	82	58	86	60	91	72	76	50	91	70	89	64	89	69	90	71	80	52	88	63
September	86	53	90	56	87	54	89	51	94	70	93	64	87	58	84	55	84	52	78	42
October	81	38	83	25	84	37	84	29	90	37	87	34	85	36	91	48	84	39	80	32
November	91	51	88	35	90	33	91	41	86	36	92	35	92	38	92	41	84	32	92	42
December	95	46	92	39	94	48	90	37	94	47	95	43	94	57	94	51	96	61	96	46

M: Morning; E: Evening.

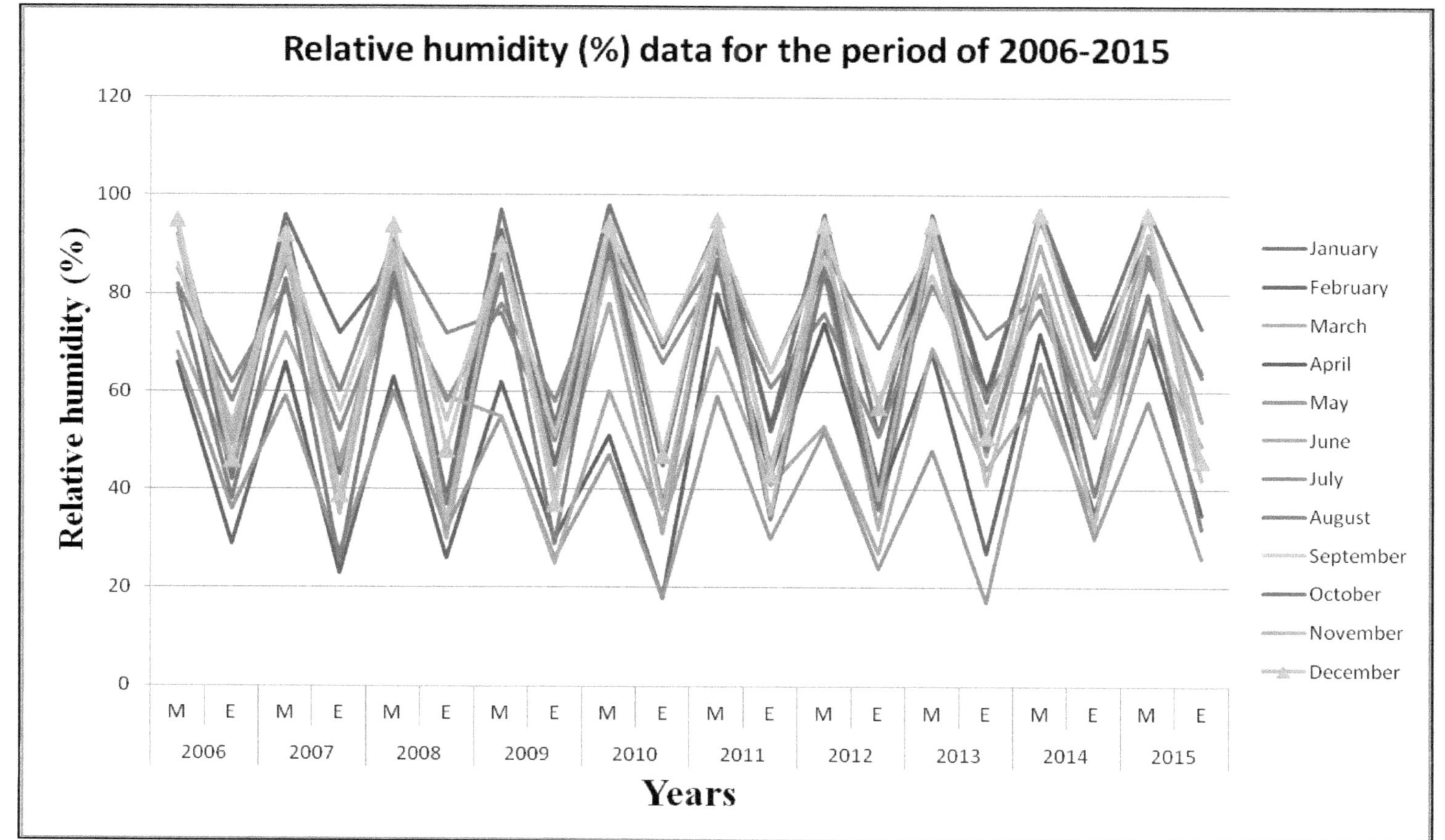
Relative humidity (%) data for the period of 2006-2015
Relative humidity (%)
120
100
80
60
40
20
0
M E M E M E M E M E M E M E M E M E M E
2006 2007 2008 2009 2010 2011 2012 2013 2014 2015
Years
January
February
March
April
May
June
July
August
September
October
November
December

months. Mean morning relative humidity remains around 80 to 90 per cent from July to end of March but decreases to about 40 to 50 per cent by April to June end.

Table 4: Monthly Climatic Normals at Hisar (50 years)

Month	*Parameters*				
	Maximum Temperature (°C)	*Minimum Temperature (°C)*	*Sunshine (Hours)*	*Evaporation (mm)*	*Rainfall (mm)*
January	20.1	4.5	7.3	2.0	11.9
February	20.7	6.5	8.0	3.1	14.6
March	28.6	10.9	8.5	8.9	19.8
April	40.9	22.3	8.4	10.4	22.5
May	40.3	22.3	8.7	12.4	22.5
June	36.3	25.7	6.4	11.6	44.9
July	36.3	25.6	5.0	8.8	134.0
August	38.1	24.0	7.0	5.8	124.7
September	38.4	21.8	8.7	6.0	66.5
October	33.5	14.8	9.3	5.1	10.5
November	28.0	9.6	8.8	5.1	10.5
December	22.6	5.4	7.3	2.3	6.3

Monthly climate normals at Hisar over last 50 years (Table 5) show that maximum and minimum temperature was recorded in April-May and December-January months. The sunshine recorded maximum hours during April-May and October-November months. Evaporation in open pen evaporimeter recorded over last 50 years shows maximum during May-June months. About 2/3rd rainfall received in this region through south-west monsoon during the months of July to September over the last 50 years.

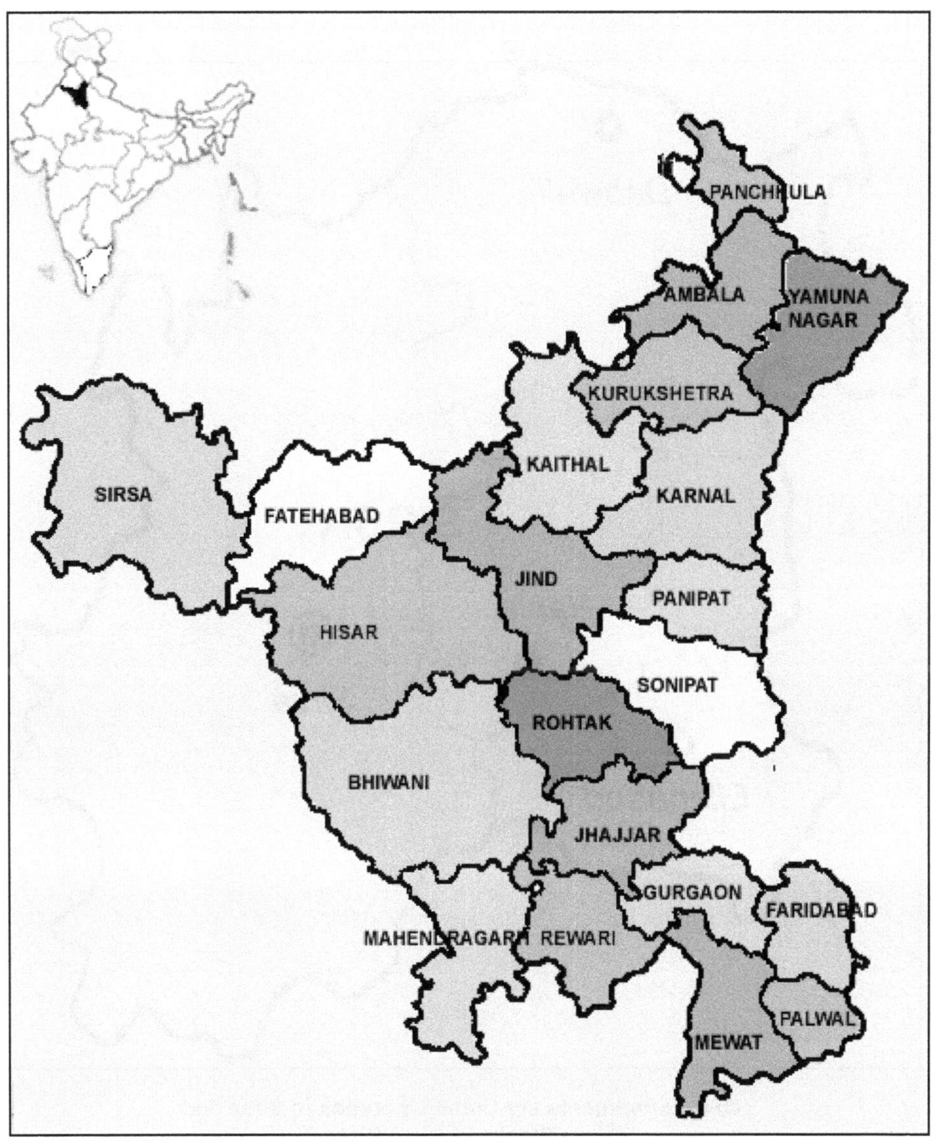

Map of Haryana (Not to the Scale).

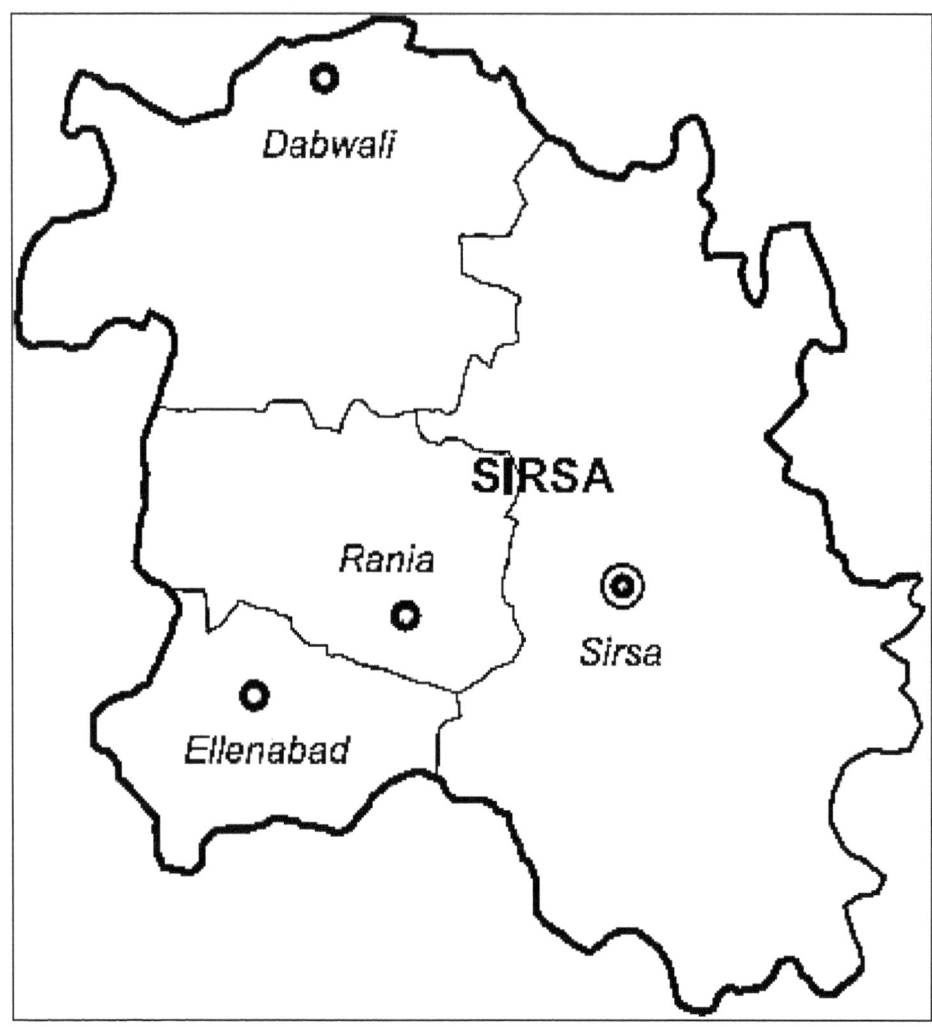

OFR Experiments are Under Operation in Sirsa and Rania Blocks of Sirsa District.

Chapter 4

Integrated Farming System Model (On Station Trial)

Diversified Farming System Model: An agro-ecological Cropping System Based Alternative for Enhancement of Rural Livelihood for Small and Marginal Farmers

Agriculture is the main enterprise of India which provides approximately 65 per cent employment to the population of the country and it also contributes about 15 per cent of gross domestic product (GDP). Due to urbanization, burgeoning population and subsequent fragmentation of land holding, pressure on available cultivating land is increasing day by day as well as uninterrupted rice-wheat rotation also led to over exploitation of natural resources and thereby less productivity. In view of the above facts, there is sturdy need to commercialize and diversify agriculture and to ensure on all sides of development of farming families. Farming should be considered as a system in which crop and other allied enterprises that should be compatible and complimentary to each other and can be combined together. It should include all components of land such as soil, water, crop, livestock, labour and other resources. The study of farming system and application of farming system approaches can bring a ray of hope for the betterment of farmers.

In India majority of farmers have holding less than two hectares which come under the category of small land holder. Per capita availability of land for agriculture purpose has declined from 0.48 ha in 1951 to 0.15 ha at present and is expected to decline to 0.09 ha by 2050 AD. In Haryana also, majority of the farmers fall in the category of small and medium land holders. The problems encountered by small and medium land holders are entirely different from large land holders. Small farmers need to produce continuous, reliable and balanced supply of food as well as cash for basic needs and recurrent farm expenditure. The small land holder farmer needs multiple enterprises for livelihood. Producing higher and quality food from small land holding warrants maintaining good soil health and other resource base. This is important in the light of the fact that most of our soils have deteriorated through indiscriminate use of chemicals.

With the decline in land for agriculture, efforts are to be made for increasing productivity of different components of farming system like crop, dairy, poultry, piggery, goat keeping, duck keeping, apiculture, sericulture, vegetable production, horticulture, mushroom culture etc. A farmer depends upon his/her family requirement on this multi-enterprise system. However, to have a systematic integration of multi-enterprise system in a scientific manner, components need to be chosen in such a manner that product or by-product of one component becomes the input for other component and thus becoming complimentary and organically interlinked to each other without wastage. For example, by-product and waste material from crops may be used as raw material for many industrial uses or as fuel and feed/fodder for cattle. Animal dung may be used as fuel, urine and crop residue may be used for preparation of compost/FYM which in turn can be used by crops. Through organic sources in the system about 25 per cent of recommended NPK for crops, can be substituted by by-product of different components and crop residue can substitute 12 per cent of recommended NPK.

The average farm holding size of Haryana was found to be 1.15 ha with 7 family members (5 adults and 2 children). The analysis of existing farming system indicates presence of crop and livestock components. Further detailed analysis of cropping systems practiced by the farmers indicates that rice – wheat and cotton – wheat as a major cropping system grown in north- east and south-west zone of the state followed by pearl millet-wheat and other crops such as fodder, vegetables, fruits. In the livestock, two buffalos and one cattle was found to be present. There were no complimentary and supplementary enterprises in the existing systems. The existing system was found to fully meet the cereals and milk requirement of family as per ICMR standard and dry fodder for the livestock component. However, the existing system was found to be deficient in producing the pulses, oilseeds, vegetables and fruits requirement of family. The system is also deficient of green fodder affecting the productivity of livestock component in the system. The commodities like pulses, oilseeds, fruits and vegetables are procured for family consumption at higher prices. Often, the families don't consume the required quantity of pulses, oilseeds and vegetables due to the higher market price. If these products are produced within the farm, the same will be available cheaper and the saving of money can be enhanced for the farm family.

Methodology

A farming system model was synthesized using the scientific inputs from the on-station research experiment conducted at CCS Haryana Agricultural University, Hisar to improve the productivity and profitability of the farm. Accordingly, the cropping systems were modified by including cereals, pulses, oilseeds, vegetables and fruits to meet the family needs. The area allocation was also made accordingly.

The synthesized cropping systems included cotton-wheat (0.20 ha), moongbean-wheat (tall) (0.05 ha), pearl millet - mustard (0.20 ha), sorghum (F)-wheat (0.20 ha), sorghum (F)-berseem (F) (0.075 ha), sorghum (F)-oats (F) (0.075 ha), cluster bean-potato -okra (0.05 ha) and okra -palak- bottle gourd (0.05 ha). Moongbean and mustard were added mainly to produce the sufficient pulses and oilseeds for the family. The livestock component of 2 buffaloes were kept as such but provision for producing sufficient green fodder was kept by including sorghum, oat and berseem in the cropping systems. In order to enhance, the income and resource recycling, complementary enterprises such as vermicompost (75 m^2) and quality FYM production (75 m^2) and karonda and bael as boundary plantation were incorporated. Karonda serves as the live fence and produces fruits which can be used for making pickles. Further, it can also protect the farm from blue bull or stray animals. Lemon + Guava + marigold/vegetable (0.06 ha) and mushroom (75 m^2) was added as income supplementing activities in the model. This model was started during 2011-12 and is continued till date at the Agronomy Research Farm of the university under AICRP on IFS. Soil of the experimental field was sandy loam having available nitrogen 152 kg ha^{-1}, phosphorus 11.5 kg ha^{-1} and potassium 289 kg ha^{-1}. The organic carbon content of the soil was 0.31 per cent. The climate of Hisar is semi-arid with an average rainfall of about 450 mm. In Hisar, about 65 per cent rain was received from south-western monsoon. Integrated farming systems were developed with an objective to fulfil the requirement of farm family and to improve the livelihood security of small and marginal farmers of irrigated area. To achieve this, the area of 9600 m^2 was allotted to various cropping systems. The cropping systems involved are cereals (pearl millet, wheat, wheat tall), pulses (mungbean), oilseed (mustard), commercial crop (cotton), green fodder (sorghum, berseem, oat), vegetables (cluster bean, potato, okra, palak, bottle gourd) and horticultural plants (lemon, guava).

Various cropping systems (Table 1) were evaluated during the period (2011-12 to 2015-16) under study. For comparison between crops sequences, return over variable cost/net return was calculated. Prevailing minimum support price/ university rates for particularly year basis were used for computing economic returns. The benefit: cost for different cropping systems was calculated dividing the return over variable cost/net return by cost of cultivation in a system. Employment generation in terms of man days engaged in a system was also recorded. The crops were raised as per the recommended package of practices. Urea and DAP were used as source of nitrogen and phosphorus, $ZnSO_4$ was used as a source of Zn and MOP was used as a source of potassium. The crops were harvested at physiological maturity. All the observations were recorded as per their area on kg/plot basis and presented as such in the final result.

A 7 member family having 5 adults and 2 children's requires 1550, 200, 130, 900, 200, 1120 and 154 kg of cereals, pulses, oilseeds, vegetables, fruits, milk and fish, respectively per annum as per ICMR standards to meet the nutritional requirement. It has been found that the synthesized model in 1.0 ha area is able to produce sufficient quantity of these produces required for the family. Apart from this, the system also generates marketable surplus of cereals, oilseeds, vegetables,

Table 1: Yield in different Cropping Systems in Integrated Farming System Model in One ha

Treatment	*Area (m^2)*	*Production (kg/plot)*														
		Kharif					*Rabi*					*Summer*				
		2011-12	*2012-13*	*2013-14*	*2014-15*	*2015-16*	*2011-12*	*2012-13*	*2013-14*	*2014-15*	*2015-16*	*2011-12*	*2012-13*	*2013-14*	*2014-15*	*2015-16*
Cotton – wheat	2000	492	518	562 (1816)	400	390	1147	1108	1248 (1372)	1000 (1200)	1200 (1300)	-	-	-	-	-
Mungbean – wheat (tall)	500	159*	106	140 (486)	10	30 (15)	144*	160	176 (342)	125 (200)	147 (220)	-	-	-	-	-
Pearl millet – mustard	2000	565	690	722 (2044)	700 (1600)	890 (1650)	473	482	510 (1574)	350 (600)	380 (625)	-	-	-	-	-
Sorghum (F) – wheat	2000	2810	2790	5886	10000	10700	1176	1188	1246 (1350)	1000 (1200)	1200 (1300)	-	-	-	-	-
Sorghum (F) – berseem (F)	750	1020	1044	2124	3750	3810	4810	4896	5012	3695	3760	-	-	-	-	-
Sorghum (F) – Oat (F)	750	990	970	1968	3750	3820	1800	1834	1896	4800	4905	-	-	-	-	-
Cluster bean-potato-okra	500	78	30	180	100 (150)	20	215	650	228	1320	849	120	46	42	144	92
Okra – palak – bottle gourd	500	201	200	36	1441	40	202	250	540	1597	35	115	30	38	-	673
Horticultural crop (lemon+ guava) +marigold/and vegetable	600	940	176	148	-		174	162	-	2.5 (Marigold seed yield)	Marrigold flower 95 Marrigold seed 1.0 Lemon 25	-	-	-	-	

*Pearl millet-wheat(tall) cropping system.

Discussion on IFS Model with Farmers during Annual Kisan Mela of the University.

Farmers Visiting IFS Experiments during Annual Kisan Mela of the University.

IAS Probation Officers Visit to IFS Model.

Dr. J.S. Sandhu, DDG Visit to the Model along with VC.

fruits, milk and mushroom, respectively ensuring sufficient income for the family besides improving the availability of these products in the market.

In farming system, the farm is viewed in a holistic manner. A combination of one or more enterprises with cropping when carefully chosen, planned and executed gives greater dividends than a single enterprise, especially for small and marginal farmers. Farm as a unit is to be considered and planned for effective integration of the enterprises to be combined with crop production activity, such that the end product of waste of one enterprise are utilized effectively as input in other enterprise. For example, the waste of dairying *viz.*, dung, urine, refuse etc are used in the preparation of vermicompost or FYM or compost which serves as an input in cropping system. Likewise, the straw obtained by the crops is used as fodder for dairy cattle. The selection of enterprises must be based on the cardinal principle of minimizing the competition and maximizing the complementariness between the enterprises. This is a highly location specific task but it will also ensure household nutritional security. Such efforts may take time and once created the long term goal of the nation to reduce poverty, unemployment and malnutrition can be achieved. In the present experiment the above aspect is very well being explained and defined in Figure of the model at page 32.

Productivity under different Cropping Systems

Cotton-wheat is an important cropping system of Haryana. On an area of 2000 m^2 seed cotton yield of 492, 518, 562, 400 and 390 kg and wheat yield of 1147, 1108, 1248, 1000 and 1200 kg was recorded in 2011-12, 2012-13, 2013-14, 2014-15 and 2015-16, respectively (Table 1). The yield of mungbean-wheat (tall), pearl millet-mustard, sorghum (F)-wheat, sorghum (F)- berseem (F), sorghum (F)-oat (F), cluster bean-potato-okra, okra-palak-bottle gourd and inter cropping of different field crops, vegetables, flowers etc. under horticulture plants as per their area allotted, are shown in Table 1 for all the years under study since 2011-12.

Among the fodder based cropping systems, sorghum-barseem produced maximum fodder during different year of study as compared to sorghum (f)-wheat and sorghum-oat cropping system. Two vegetable based cropping systems were included in the model and were evaluated based on their productivity and economic returns. In cluster bean-potato-okra cropping system yield of cluster bean to the tune of 78, 30, 180,100 and 20 kg/plot and of potato during *Rabi* 215, 650, 228, 1320 and 849 kg/plot was harvested during 2011-12, 2012-13, 2013-14, 2014-15 and 2015-16, respectively. During summer, okra produced 120, 46, 42, 144 and 92 kg/plot yield during different year of study.

Lemon and guava were taken as horticulture crops in 600 m^2 area. In the initial two years cotton during *Kharif* and wheat during *Rabi* season were taken as intercrops, which gave 940,176 kg/plot of cotton and 174,162 kg/plot yield of wheat. In the last two years marigold was taken as intercrop in the horticulture area. Marigold produced 2.5 kg seed during 2014-15 and 95 kg flowers and 1.0 kg seed was produced during the year 2015-16. In addition, 25 kg lemon were also harvested during the year 2015-16.

Cropping System in IFS Model at CCSHAU, Hisar.

Cropping System in IFS Model at CCSHAU, Hisar.

Economic Returns

Highest gross return as well as return over variable cost/net return were obtained in 2000 m^2 area under cotton-wheat cropping system followed by pearl millet-mustard and sorghum (F)-wheat during all five years of study (Tables 2 and 3). It was due to high cost of cotton crop. In the year 2015-16 highest gross (Rs. 54860) and return over variable cost/net returns (Rs 45470) were recorded in sorghum (F)-wheat cropping system. It was followed by cotton-wheat cropping system with gross return of Rs. 42370 and Rs. 28014, respectively. The cropping system with both crops as green fodder gave gross return of Rs. 21948 and Rs. 24430 per plot of 750 m^2 and return over variable cost/net return of Rs. 19504 and Rs. 22019 during the year 2015-16. Among the two vegetable based cropping systems potato growing was more remunerative as compared to palak in *Rabi* season. The return over variable cost/net return recorded during the year 2015-16 in cluster bean-potato-okra and okra-palak-bottle gourd were Rs. 6371 and 5709, respectively in an area of 500 m^2. In the horticultural orchard lemon and guava were planted in 600 m^2 area. As these plants are still to bear fruits, intercrops were taken in between. In the first two years cotton and wheat were taken. During 2014-15 and 2015-16 marigold was intercropped and it was found more remunerative. A net income of Rs. 8650 and 10789 was obtained by intercropping marigold in lemon and guava during the year 2014-15 and 2015-16, respectively.

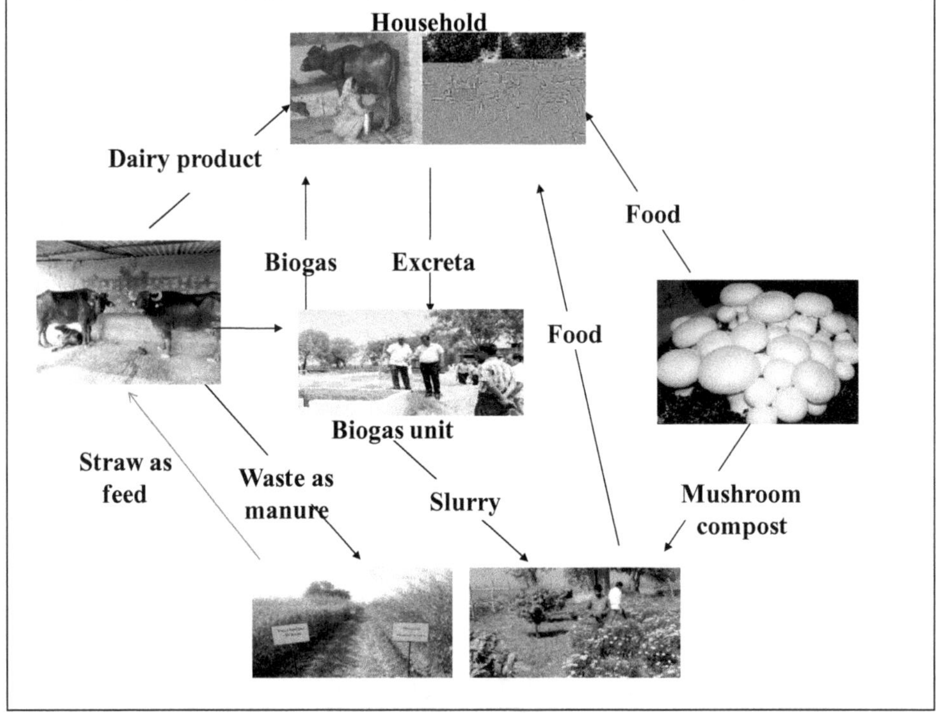

Crop-Livestock-Mushroom Farming System Model at Hisar.

Table 2: Gross Returns and Cost of Cultivation in different Cropping Systems in Integrated Farming System Model in One ha^{-1}

Treatments	*Gross Returns (Rs./plot)*					*Cost of Cultivation*				
	2011-12	*2012-13*	*2013-14*	*2014-15*	*2015-16*	*2011-12*	*2012-13*	*2013-14*	*2014-15*	2015-16
Cotton – wheat	28505	35344	41486	37400	42370	12770	12736	4250	11411	14356
Mungbean – wheat (tall)	3851	7224	11845	3800	6630	1760	2194	2060	2014	3207
Pearl millet – mustard	18254	25575	28905	27950	32905	7120	7622	3190	5551	6226
Sorghum (F) – wheat	18722	20582	26792	45600	54860	7860	5936	3875	7565	9390
Sorghum (F) – berseem (F)	7169	5346	10385	18611	21948	2641	720	2050	2175	2444
Sorghum (F) – Oat (F)	3526	2524	5217	21375	24430	2453	614	1875	2142	2411
Cluster bean-potato-okra	3423	8550	7650	20880	12770	2499	1540	2010	6611	6399
Okra – palak – bottle gourd	3859	6800	13616	6014	11730	2561	3340	2455	5701	6021
Horticultural crop (lemon+ guava) +marigold/and vegetable	3157	5058	6020	9889	12200	1227	2635	1005	1239	1411
Dairy component + FYM	-	-	27454	70500	42566	-	-	18000	23605	22320
Mushroom	-	-	-	4592	5850	-	-	-	992	995
Recycled (Vermicompost/Agri. waste)	-	-	1400	18300	28120	-	-	-	-	-
Total	90466	117003	180770	284911	296379	40891	37337	40770	69006	75180

Table 3: Return Over Variable Cost/Net Returns in different Cropping Systems in Integrated Farming System Model in One ha^{-1}

Treatments	*Return Over Variable Cost/Net Returns (Rs./plot)*					*Percent Contribution*				
	2011-12	*2012-13*	*2013-14*	*2014-15*	*2015-16*	*2011-12*	*2012-13*	*2013-14*	*2014-15*	*2015-16*
Cotton – wheat	15735	22608	37236	25989	28014	31.73	28	28.72	17.67	16.67
Mungbean – wheat (tall)	2091	5030	9785	1786	3423	4.21	6	7.93	1.21	2.03
Pearl millet – mustard	11134	17953	25715	22399	26679	22.26	23	19.83	15.22	15.88
Sorghum (F) – wheat	10862	14646	22917	38035	45470	21.72	18	17.68	25.85	27.06
Sorghum (F) – berseem (F)	4528	4626	8335	16436	19504	9.05	6	6.43	11.18	11.61
Sorghum (F) – Oat (F)	1073	1910	3342	19233	22019	2.15	3	2.58	13.07	13.10
Cluster bean-potato-okra	924	7010	5640	313	6371	1.84	9	8.60	0.21	3.79
Okra – palak – bottle gourd	1298	3460	11161	14269	5709	2.59	4	4.35	9.69	3.39
Horticultural crop (lemon+ guava) +Marigold/and Vegetable	1930	2423	5015	8650	10789	3.86	3	3.87	5.87	6.42
Dairy component	-	-	9454	46895	20246	-	-	6.75	21.72	9.15
Mushroom	-	-	-	3600	4855	-	-	-	1.66	2.19
Recycled (Vermicompost/FYM/ Agri. waste)	-	-	1400	18300	28120	-	-	1	8.47	12.71
Total	49575	79,666	1,40,000	2,15,905	2,21,199	-	-	-	-	-

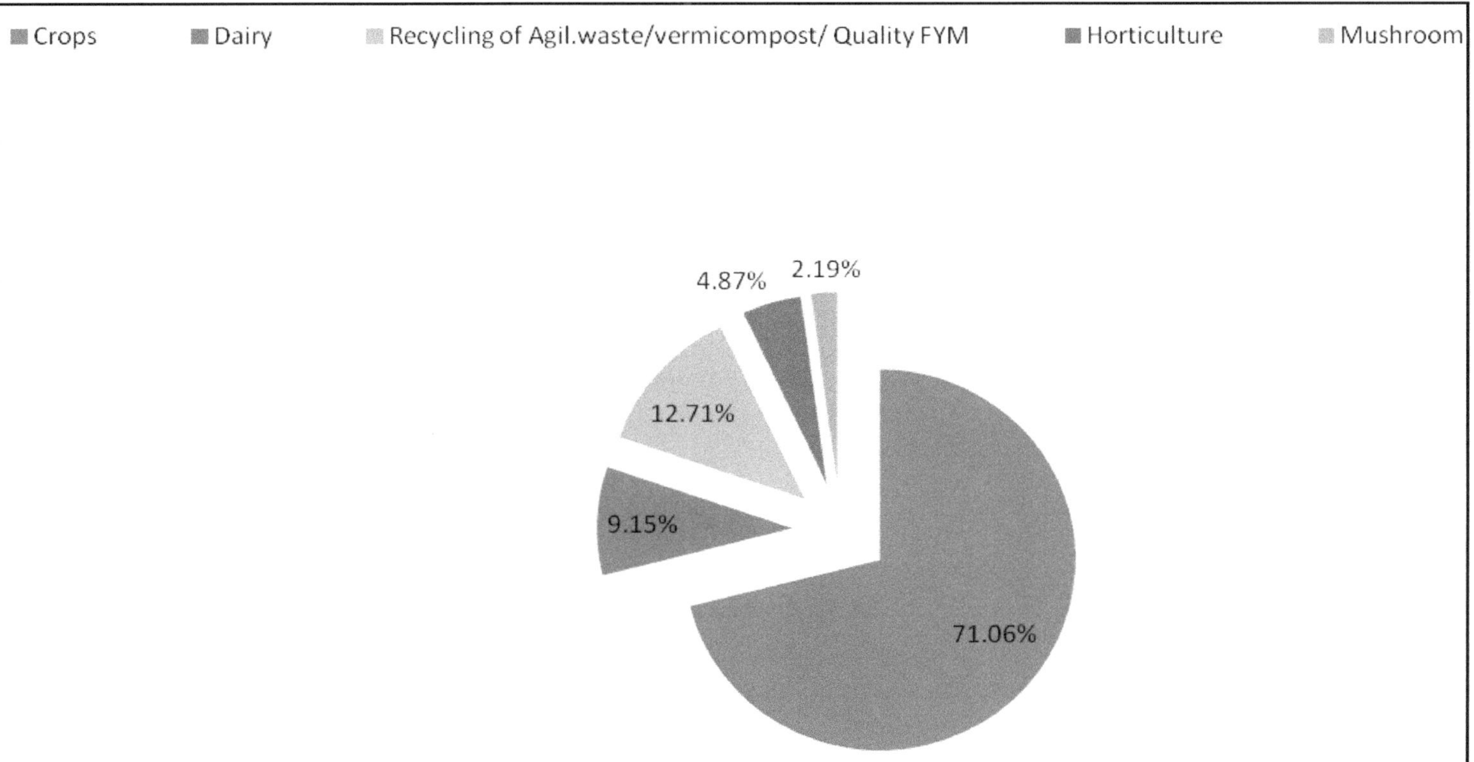

Figure 1: FS-1: Per cent Contribution of different Components during the Year 2015-16 in the IFS Model.

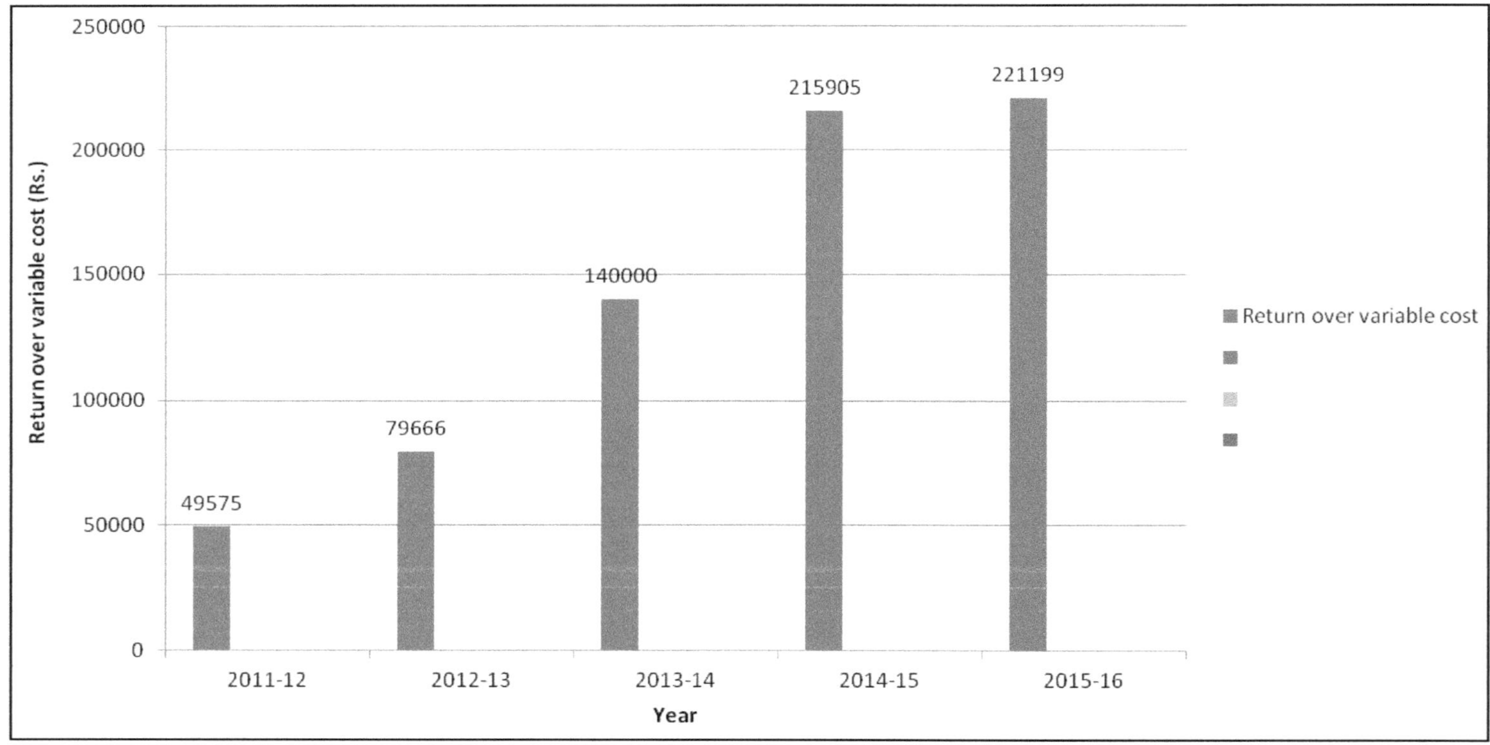

Figure 2: Return over Variable Cost/Net Returns (Rs.) in the IFS Model.

During the initial two years only different cropping system were tried in the model, which gave a return over variable cost/net return of Rs. 49575 and Rs. 79666 during 2011-12 and 2012-13, respectively. During third year, dairy (Two buffalos) and vermicompost were added to the model and thus the net income was enhanced to Rs. 140000. During the year 2014-15, mushroom and floriculture were added to the model, which further raised the net income to the tune of 2,15,905. In the recent year *i.e.* 2015-16 a net income of Rs. 2,21,199 was achieved from the model. Among this, Rs. 145109 were obtained from various cropping systems, and Rs. 12080 from two vegetable based cropping system were earned. Intercropping of marigold was found more remunerative and gave Rs. 10789. One buffalo was dry during the entire year. The net income of Rs. 20246 was obtained from dairy component from another buffalo. Rs. 4855 and Rs. 28120 were obtained from mushroom production and recycling (Vermicompost/Agricultural waste/Quality FYM), respectively. Among all the five years highest return over variable cost/net return of Rs. 2,21,199 was recorded from one hectare during the year 2015-16.

Percent Contribution

The various cropping systems included in the model contributed maximum to the net income during all the year of study. Among the cropping systems cotton-wheat contributed maximum *i.e.*31.73, 28.0, 28.72 per cent during 2011-12,2012-13 and 2013-14, respectively (Table 3). Fodder based cropping system (Sorghum (F)-wheat) contributed maximum *i.e.* 25.85 and 27.06 per cent during 2014-15 and 2015-16, respectively. The six cereal and fodder based cropping system contributed maximum to the net income during all the year of study. The percent contribution was to the tune of 91.12, 84.0, 83.17, 84.20 and 86.35 per cent during 2011-12,2012-13,2013-14,2014-15 and 2015-16, respectively. The two vegetable based cropping systems tried contributed 4.40, 13.0, 12.95, 9.90 and 7.18 per cent during 2011-12,2012-13,2013-14,2014-15 and 2015-16, respectively. The other major contributor to the model was dairy component. The two buffalos under dairy component were added during 2013-14 in the model. The percent contribution from this component in the return over variable cost/net returns was recorded as 6.75, 21.72 and 9.15 per cent during 2013-14, 2014-15 and 2015-16, respectively. The recycling of resources within the model is another aspect contributing to the net income. 8.47 and 12.71 per cent was contributed to the net income during 2014-15 and 2015-16, respectively from recycling of the resources *i.e.* vermicompost/FYM/Agri. waste.

Employment Generation

One of the important objectives of the Integrated Farming system is to provide employment to the farming family round the year. In the present model highest employment was generated in the year 2015-16 (340 man days). While during 2014-15 the employment was generated to the tune of 257 man days. Further it has been seen that dairy component generate maximum employment round the year during both the years.

Farming System Diversification/Intensification

The increased cropping intensity will be a key strategy for future gains in crop production. Short duration pulses, oilseeds and other high value crops will find their definite niche as sequential or intercrops, rather than replacing the major cereal crops having higher yield stability. Hence, an increased cropping intensity will contribute substantially to additional demands of food and cash crops. Development of new crop varieties with more efficient photosynthetic apparatus and shorter duration would be of immense help in increasing cropping intensity. Similarly, bio-intensive diversified complementary cropping systems would enable small and marginal farmers to utilize limited land and water resources in more efficient manner. The diversified cropping systems need to be considered in farming system perspective. The land based enterprises such as dairy, mushroom, vermicompost, FYM etc. were included to the model to complement the cropping programme to get more income and employment for small farmers of Haryana. A return over variable cost/net returns of Rs 221199 can be realized with an investment of Rs 75180 in 1.00 ha area which also generated 340 man days of employment with a resource use efficiency of Rs 3.94/Re invested thus ensuring the livelihood of small and marginal farmer of the state (Table 4).

Table 4: Crop + Livestock + Mushroom based Integrated Farming System Model for Increasing Income and Employment of Small and Marginal Farmers (1.0 ha) at Hisar during the Year 2015-16

Parameters	*Total Expenditure (Rs)*	*Return Over Variable Cost/Net Returns (Rs)*	*Return Re-invested*	*Employment Generation (Man days)*
Field crops	23789	58116	3.44	130
Fodders	14245	86993	7.10	
Vegetables	12420	12080	1.97	
Horticulture	1411	10789	10.68	10
Dairy component	22320	20246	1.90	180
Mushroom	995	4855	5.87	05
Recycled (Vermicompost/ Agri. waste	-	28120	-	-
Total	75180	221199	3.94	340

Conclusion

Diversification of existing farming systems clearly demonstrated the advantages. It has been observed that productivity gain of 2 to 3 times and increase in return over variable cost/net returns of 3 to 5 times is possible with improved systems. Further, resource saving of 40 to 50 per cent can also be ensured besides enhancing the income of household to the level of at least Rs 2,21,199 per annum. Additional employment generation of 70 to 80 per cent is also possible. Improved diversified systems also ensure household nutritional security.

Way Forward in Farming System Diversification

Focus should be given mainly on family and market-oriented diversification and livelihood improvement by considering the options of alternative cropping, novel livestock systems and adding of value to primary (raw) products. Market-family driven diversification options can be achieved through, capacity building, crop and forage rotation, including perception based Desi/Shaiwal cows, small ruminants (Goats), poultry birds and ensuring monthly income flows through product diversification.

Acknowledgement

The authors feel highly indebted to the Indian Institute of Farming Systems Research, ICAR, Modipuram for providing technical guidance and financial support to carry out this experiment for such a long time (since 2011-12) in the Department of Agronomy, CCS Haryana Agricultural University, Hisar. The authors also thankfully acknowledge the contribution of scientific staff worked in the project since 2011-12 to till date.

Chapter 5

Cropping Systems

Identification of Need Based Cropping System for different Agro-climatic Regions of Haryana

Our analysis of land use for the last 40 years in the four states suggests that Punjab and Haryana have reached the absolute limit of expansion of area under cultivation with almost 84 per cent of the area being cultivated. Six of 8 per cent of the area of these states is under urban uses. Another 5 per cent are under forests (mainly strip forests) and the remaining 2-3 per cent is roads, canals and other infrastructural and industrial uses. Cultivable waste as a category has virtually disappeared in these two states. Such intensive land use for agriculture is sustainable only with increasing and continued high doses of balanced nutrients and other inputs such as chemical fertilizers and insecticides. The proportion of area available for cultivation in Himachal Pradesh and Jammu and Kashmir because of topography and physiography is rather small and cannot be expanded without major private and public investment that in return will result in major ecological problems and should be avoided.

The cropping pattern in the region has under gone a substantial change, with wheat and rice emerging as a major crop rotation in Punjab and half of Haryana. Its expansion in Himachal Pardesh and Jammu and Kashmir has been moderate. Crops that have been replaced by wheat and rice, are gram, bajra, barley, millets, and pulses. Area under cotton has grown in Haryana. In absence of expansion of the sugar Industry, the area under sugarcane has remained static. The cropping pattern of the region has unnecessarily become energy intensive and is affective the static balance of the under ground water resources in the plains of Punjab and Haryana. The growth of infrastructure irrigation, and other technological factors are responsible for a major shift in cropping pattern in favour of wheat and rice in the states of Punjab and Haryana.

A number of policy steps must be taken to encourage farmers to switch from rice, which is a water and fertilizer intensive crop in the region, to crops that demand less water. This can be achieved through price policy research and development efforts, and establishment of agro processing industries as so to make sustainable alternatives more attractive to the individual farmers. Without supplementary organic manure, intensive agriculture leads to depletion of soil fertility. Ludhiana district in the green revolution state of Punjab, which records the highest yields of many crops, now also records the highest deficiencies of plants micronutrients. Extensive use of organic manure is the only way to over come the deficiency in Punjab, above 5 million tonnes of rice straw is being burnt every year during October to December, If crop residues were ploughed back into the soil, the rate of micronutrient depletion would be substantially reduced.

Turning now to the socio-economic and environmental consequences of crop pattern changes, the Green Revolution technologies have fomented, among other things, an increasing tendency towards crop specialization and commercialization of agriculture. While these developments have positive effects on land/labour productivity and net farm income, they have also endangered a number of undesirable side effects like reduced farm employment and crop imbalances. Although the expansion of commercialized agriculture has fomented new sets of rural non-farm activities and strengthened the rural-urban growth linkages, it has also weakened the traditional inter-sectoral linkages between the crop and livestock sectors. Besides, crop pattern changes also lead to serious environmental consequences that take such forms as groundwater depletion, soil fertility loss and waterlogging and salinity - all of which can reduce the productive capacity and growth potential of agriculture over the long-term. A classical example is the rice-wheat system in north western India replacing traditional crops like pulses, oilseeds and cotton.

Efficient cropping system for a particular farm depends on farm resources, farm enterprise and farm technology. The farm resources include land, labour, water, capital, and infrastructure. When land is limited, intensive cropping is adopted to fully utilize available water and labour. When sufficient and cheap labour is available, vegetable crops also include in the cropping system as they require more labour. Capital intensive crops like sugarcane, banana, turmeric etc. find a place in the cropping system when capital is not a constraint. In low rainfall regions (<750mm/annum) monocropping is followed and when rainfall is more than 750 mm intercropping is practiced. With sufficient irrigation water, triple and quadruple cropping is adopted. When other climatic factors are not limiting, farm enterprises like dairying, poultry etc., also influence the type of cropping system. When the farm enterprise includes dairy, the cropping system should contain fodder crops as components. Change in cropping system takes place with the development of technology. The feasibility of growing four crop sequence in gangetic alluvial plains gave impetus to multiple cropping.

Different cropping system for the two zones of Haryana were tried in this experiment with following objective:

☆ To identify appropriate cropping systems with high productivity to suit the specific needs of different agro-climatic regions.

Year of start: *Kharif* 1987, Revised in *Kharif* 2000 and again revised in *Kharif* 2006.

Methodology

The experiment conducted from 2006-07 to 2015-16 with various cropping systems keeping in view the diversification issues of Haryana is presented here and the details are given below:

Table 1: Details of Crop/Varieties and NPK Doses (kg ha^{-1}) in different Cropping Systems

Treatments	*Kharif*	*Rabi*	*Summer*
T_1	Pearl millet (HHB 197) (125-62.5-0)	Wheat (WH-711) (150-60-0)	-
T_2	Cotton (H-1226) (175-60-60)	Wheat (WH-711) (150-60-0)	-
T_3	Pearl millet (HHB-197) (125-62.5-0)	Barely (BG-393) (60-30-0)	Mungbean (Satya) (20-40-0)
T_4	Cluster bean (HG-365) (20-40-0)	Broccoli (CBH-1) (100-50-0)	Onion (Hisar-1) (125-50-25) + FYM 25T
T_5	Mungbean (Satya) (20-40-0)	Mustard (RH-30) (80-30-0) + *Kashni*	-
T_6	Pearl millet (HHB-197) (125-62.5-0)	Wheat (*Desi*) (C-306) (60-30-0)	Cowpea (Veg+Residue) (Pusa Komal) (25-62.5-0)
T_7	Pearl millet (HHB-197) (125-62.5-0) + Mungbean (Satya)	Wheat (WH-711) (150-30-0)+ Mustard (RH-30)	-

Experimental design	:	BIBD
Replication	:	4
Plot size	:	10 X 8.1 m^2

Experimental Results and Research Achievements

Yield

The yield data has been presented in Table 2. The highest yield of pearl millet 3894, 3712, 3244, 3026, 3332, 3380, 3060 and 3321 was recorded in pearl millet-wheat (*desi*)-cowpea cropping system during 2008-09, 2009-10, 2011-12, 2012-13, 2013-14,

Overview of Different Cropping Systems.

Overview of Different Cropping Systems.

Overview of Different Cropping Systems.

Overview of Different Cropping Systems.

Table 2: Yield of *Kharif* Season in different Cropping Systems for the Period of 2006-07 to 2015-16

Treat-ment	*Cropping Systems*	*Kharif Yield (kg ha^{-1})*										
		2006-07	*2007-08*	*2008-09*	*2009-10*	*2010-11*	*2011-12*	*2012-13*	*2013-14*	*2014-15*	*2015-16*	*Mean*
T_1	Pearl millet – wheat	3304	3395	3681	3594	2704	3016	2832	2894	2978	2680	3108
T_2	Cotton – wheat	2861	2897	2951	2441	2184	2582	2376	2542	2494	2228	2556
T_3	Pearl millet – barley – mungbean	3340	3492	3778	3658	2826	3163	2946	3165	3216	2927	3251
T_4	Clusterbean – broccoli – onion	1722	1686	1748	1532	1018	1076	897	1006	984	0	1167
T_5	Mungbean – mustard + kashni	1249	1270	1186	1094	866	988	863	968	1034	0	952
T_6	Pearl millet – wheat (*desi*) – cowpea	3294	3476	3894	3712	2796	3244	3026	3332	3380	3060	3321
T_7	Pearl millet + mungbean – wheat + mustard	2625 + 102	2578 + 119	2714 + 198	2576 + 172	1782 + 168	2467 + 206	2347 + 182	2590 + 194	2618 + 207	2368 + 0	2467 + 155

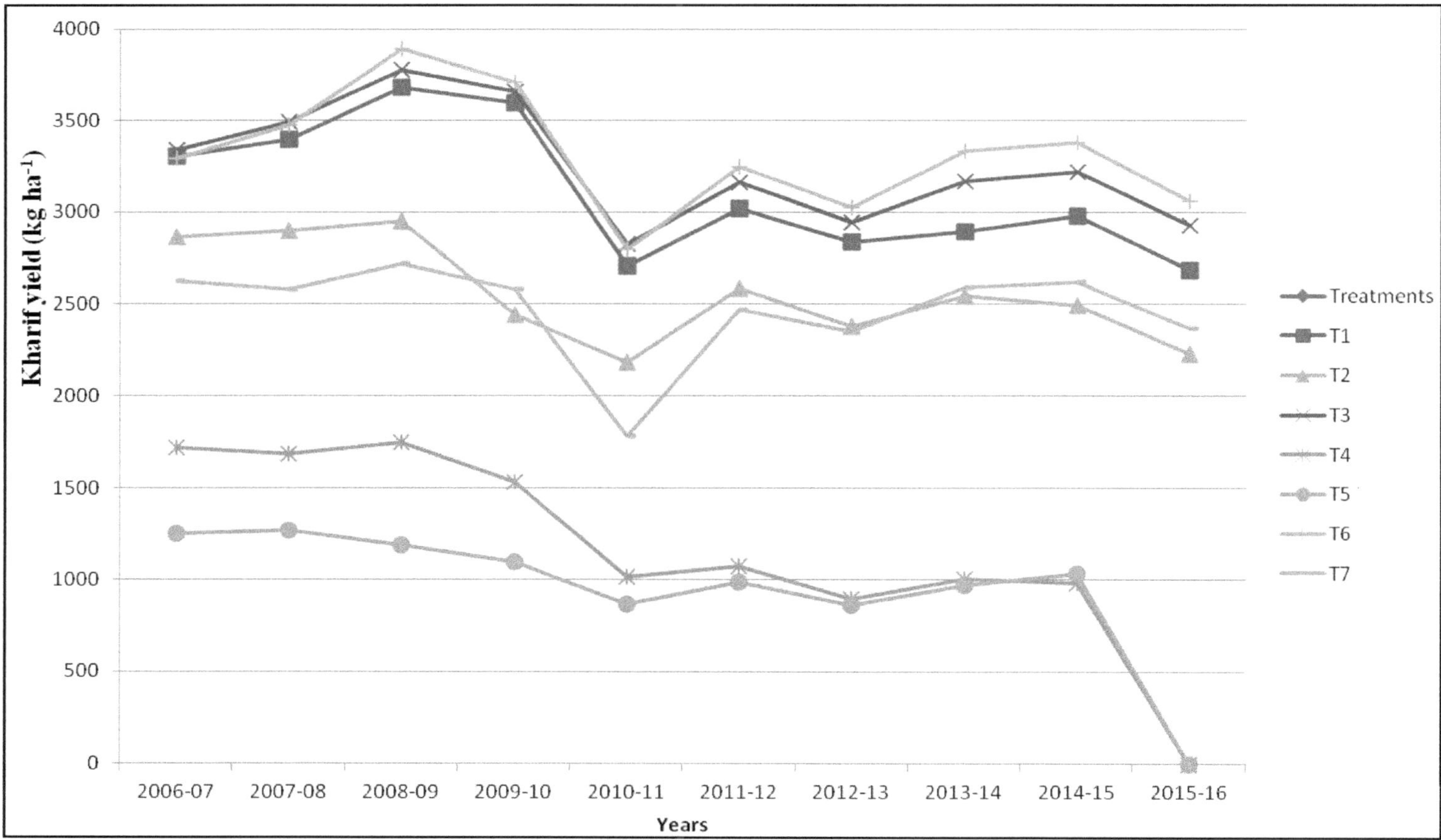

Figure 1: Yield of *Kharif* Season in different Cropping Systems for the Period of 2006-07 to 2015-16.

2014-15, 2015-16 and mean, respectively. It might be due to residual effect of cowpea crop taken in summer season. During the year 2006-07, 2007-08 and 2010-11 the highest pearl millet yield of 3340, 3492 and 2826, respectively was recorded in pearl millet-barley-mungbean cropping system.

The *Rabi* crops yield data indicate (Table 3) that wheat yield in cotton-wheat cropping system was highest 5927 and 5918 kg ha^{-1} during the years 2014-15 and 2015-16, respectively. Whereas, the wheat yield in pearl millet-wheat cropping system was highest 5060, 5685, 5948, 5462, 5464, 5634, 5266, 5586 and 5533 during 2006-07, 2007-08, 2008-09, 2010-11, 2011-12, 2012-13, 2013-14 and mean, respectively.

Summer crops *viz.* mungbean and cowpea were taken in pearl millet-barley-mungbean and pearl millet-wheat (*desi*)-cowpea cropping systems. The yield of summer crops was 702, 625, 418, 568, 396, 378, 356, 308, 324, 274 and 435 kg/ha of mungbean in pearl millet-barley-mungbean during 2006-07, 2007-08, 2008-09, 2010-11, 2011-12, 2012-13, 2013-14, 2014-15, 2015-16 and mean, respectively (Table 4). The yield of cowpea in pearl millet-wheat (*desi*)-cowpea cropping systems was 254, 813, 1235, 540, 864, 923, 648, 341, 496, 304 and 642 kg/ha during 2006-07, 2007-08, 2008-09, 2010-11, 2011-12, 2012-13, 2013-14, 2014-15, 2015-16 and mean, respectively. The yield of onion in cluster bean-broccoli-onion cropping system was 10864, 688, 9876, 6142, 21300, 2308, 2382, 4276, 5349, 5422 and 6861 in different years of study and mean, respectively. Among the summer crops, vegetable based cropping system gave the highest mean yield during all the year of study.

Wheat Equivalent Yield

Due to differential nature of crop yield and the differences in price realization, the wheat equivalent yield of different cropping systems was calculated. The highest wheat equivalent yield of 10878, 10651, 13879, 10849, 10009, 11187, 15927, 13044, 12607, 11908 and 12094 kg ha^{-1} was recorded in cotton–wheat cropping system during 2006-07, 2007-08, 2008-09, 2010-11, 2011-12, 2012-13, 2013-14, 2014-15, 2015-16 and mean, respectively (Table 5).

Gross Returns

The mean gross returns among different cropping systems ranged markedly between Rs. 96056 and Rs. 151566 ha^{-1} (Table 6). The maximum gross return was recorded in cotton-wheat cropping system during all the year of study except during 2006-07. Pearl millet-barley-mungbean recorded maximum gross return (Rs. 105093) during the year 2006-07. The mean minimum gross returns (Rs. 96056) was realized in cluster bean- broccoli- onion cropping system.

Return over Variable Cost/Net Returns

The return over variable cost/net returns among different cropping systems ranged markedly between Rs. 36001 and Rs. 90396 ha^{-1} (Table 7). The maximum return over variable cost/net return of Rs. 60310, 91239, 64829, 77492, 85616, 90618, 117704, 120554, 139251 and 90396 ha^{-1} was recorded in cotton-wheat cropping system

Table 3: Yield of *Rabi* Season in different Cropping Systems for the Period of 2006-07 to 2015-16

Treat-ment	*Cropping Systems*	*Rabi Yield (kg ha^{-1})*										
		2006-07	*2007-08*	*2008-09*	*2009-10*	*2010-11*	*2011-12*	*2012-13*	*2013-14*	*2014-15*	*2015-16*	*Mean*
T_1	Pearl millet – wheat	5060	5685	5948	5462	5464	5634	5266	5586	5720	5507	5533
T_2	Cotton – wheat	4920	5436	5682	5308	5342	5561	5197	5512	5927	5918	5480
T_3	Pearl millet – barley – mungbean	3895	3662	3741	3418	3476	3218	2976	3016	3209	3210	3382
T_4	Clusterbean – broccoli – onion	3470	18518	4104	1635	942	1790	1581	1560	2963	2938	3950
T_5	Mungbean – mustard + kashni	2450 + 100	2068 + 150	2144 + 165	2226 + 158	1768 + 207	2376 + 149	2258 + 144	2240 + 144	2351 + 124	2387 + 222	2227 + 156
T_6	Pearl millet – wheat (*Desi*) – cowpea	2990	3084	3189	2592	2814	2944	2822	2916	2989	2972	2931
T_7	Pearl millet + mungbean – wheat + mustard	4360 + 188	4870 + 118	5298 + 126	4742 + 138	4836 + 114	4964 + 146	4706 + 142	4932 + 184	5097 + 172	5150 + 158	4896 + 149

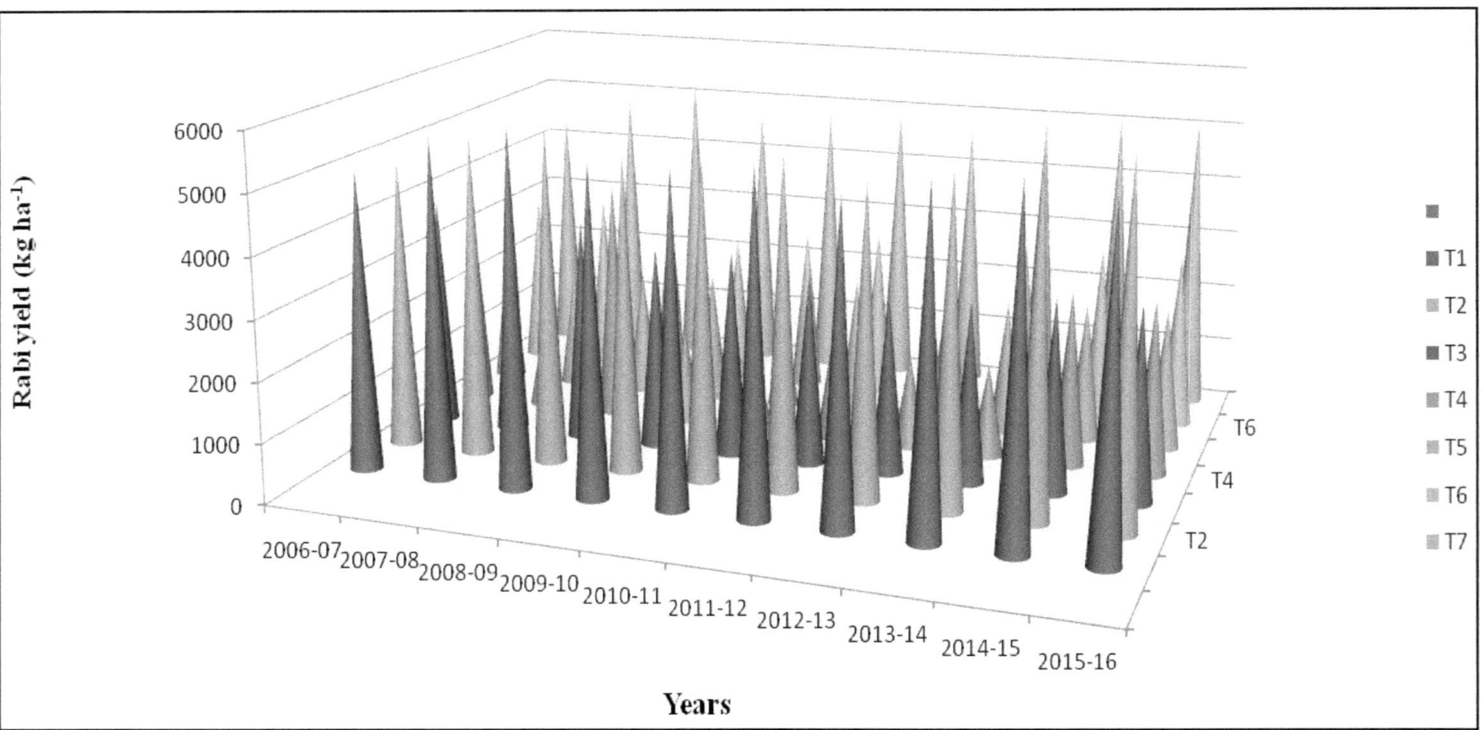

Figure 2: Yield of *Rabi* Season in different Cropping Systems for the Period of 2006-07 to 2015-16.

Table 4: Yield of Summer Season in different Cropping Systems for the Period of 2006-07 to 2015-16

Treat-ment	*Cropping Systems*	*Summer Yield (kg ha^{-1})*										
		2006-07	*2007-08*	*2008-09*	*2009-10*	*2010-11*	*2011-12*	*2012-13*	*2013-14*	*2014-15*	*2015-16*	*Mean*
T_1	Pearl millet – wheat	-	-	-	-	-	-	-	-	-	-	-
T_2	Cotton – wheat	-	-	-	-	-	-	-	-	-	-	-
T_3	Pearl millet – barley – mungbean	702	625	418	568	396	378	356	308	324	274	435
T_4	Clusterbean – broccoli – onion	10864	688	9876	6142	21300	2308	2392	4276	5349	5422	6861
T_5	Mungbean – mustard + kashni	-	-	-	-	-	-	-		-	-	-
T_6	Pearl millet – wheat (*desi*) – cowpea	254	813	1235	540	864	923	648	341	496	304	642
T_7	Pearl millet + mungbean – wheat + mustard	-	-	-	-	-	-	-	-	-	-	-

Table 5: Wheat Equivalent Yield (WEY) in different Cropping Systems for the Period of 2006-07 to 2015-16

Treat-ment	*Cropping Systems*	*Wheat Equivalent Yield (kg/ha)*										
		2006-07	*2007-08*	*2008-09*	*2009-10*	*2010-11*	*2011-12*	*2012-13*	*2013-14*	*2014-15*	*2015-16*	*Mean*
T_1	Pearl millet – wheat	7159	7722	8811	8193	7498	7982	7731	8266	8379	7748	7949
T_2	Cotton – wheat	10878	10651	13879	10849	10009	11187	15927	13044	12607	11908	12094
T_3	Pearl millet – barley – mungbean	5966	5562	7654	8785	5604	5896	5884	6012	6457	5897	6372
T_4	Clusterbean – broccoli – onion	9703	4183	9847	6349	10582	9643	7001	6987	9570	4518	7838
T_5	Mungbean – mustard + kashni	7176	6396	7167	6398	7172	8048	7895	8256	8457	5260	7223
T_6	Pearl millet – wheat (*Desi*) – cowpea	7257	6816	7495	6143	6080	6871	7679	7405	6361	5729	6784
T_7	Pearl millet + mungbean – wheat + mustard	6589	6939	8084	7371	7599	7689	7690	8386	8483	7477	7631

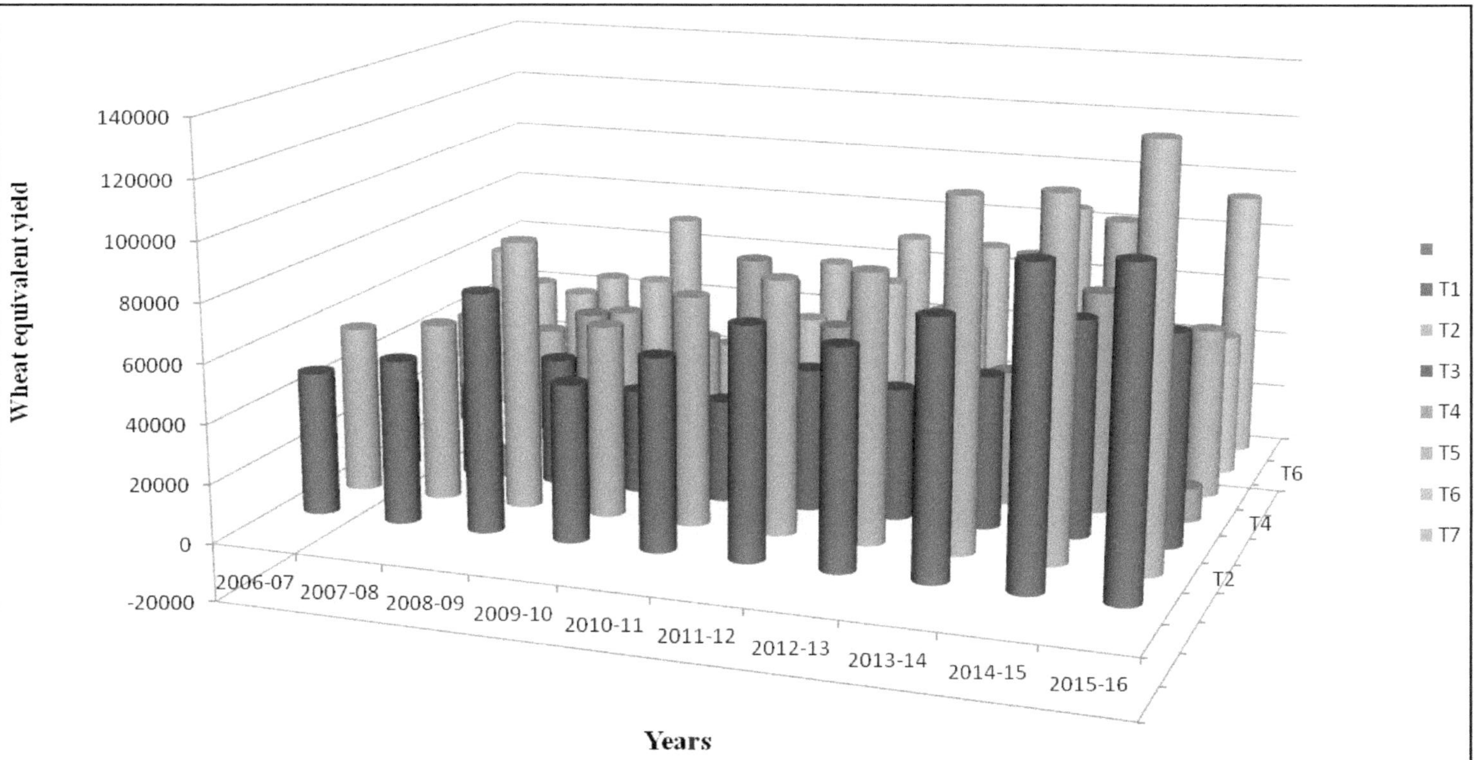

Figure 3: Wheat Equivalent Yield (WEY) in different Cropping Systems for the Period of 2006-07 to 2015-16.

Table 6: Gross Returns (Rs/ha) in different Cropping Systems for the Period of 2006-07 to 2015-16

Treat-ment	*Cropping Systems*	*Gross Returns (Rs/ha)*										
		2006-07	*2007-08*	*2008-09*	*2009-10*	*2010-11*	*2011-12*	*2012-13*	*2013-14*	*2014-15*	*2015-16*	*Mean*
T_1	Pearl millet – wheat	80857	99492	118845	111753	104325	124320	123753	133974	153149	157780	120824
T_2	Cotton – wheat	84043	118446	146850	126920	134392	154782	165319	184561	187109	213242	151566
T_3	Pearl millet – barley – mungbean	105093	76617	89395	94980	80076	96823	96952	106678	127853	132192	100666
T_4	Clusterbean – broccoli – onion	72723	42374	111647	74155	126327	121116	101814	98083	143410	68907	96056
T_5	Mungbean – mustard + kashni	84966	68348	83514	77963	95418	108392	114656	115263	129155	89616	96729
T_6	Pearl millet – wheat (*Desi*) – cowpea	67577	96304	103445	99456	86211	108723	119746	125673	121496	124188	105282
T_7	Pearl millet + mungbean – wheat + mustard	84696	88100	107752	103803	93878	117501	120438	134506	151299	146919	114889

Table 7: Return Over Variable Cost/Net Returns (Rs/ha) in different Cropping Systems for the Period of 2006-07 to 2015-16

Treat-ment	*Cropping Systems*	*Return Over Variable Cost/Net Returns (Rs/ha)*										
		2006-07	*2007-08*	*2008-09*	*2009-10*	*2010-11*	*2011-12*	*2012-13*	*2013-14*	*2014-15*	*2015-16*	*Mean*
T_1	Pearl millet – wheat	48254	55177	80449	52802	64525	77730	73437	85579	105115	107365	75043
T_2	Cotton – wheat	56351	60310	91239	64829	77492	85616	90618	117704	120554	139251	90396
T_3	Pearl millet – barley – mungbean	29460	30967	43424	35234	34676	48223	44462	51658	73286	71801	46319
T_4	Clusterbean – broccoli – onion	34195	-10058	52713	-14183	77127	56116	31614	45612	75330	11541	36001
T_5	Mungbean – mustard + kashni	39990	37308	46838	40093	62668	55202	57210	79551	93762	58162	57078
T_6	Pearl millet – wheat (*Desi*) – cowpea	58737	44790	51869	30158	42511	58523	65526	64758	61034	48733	52664
T_7	Pearl millet + mungbean – wheat + mustard	40065	44400	69356	44153	57228	68990	68048	84596	101691	92848	67138

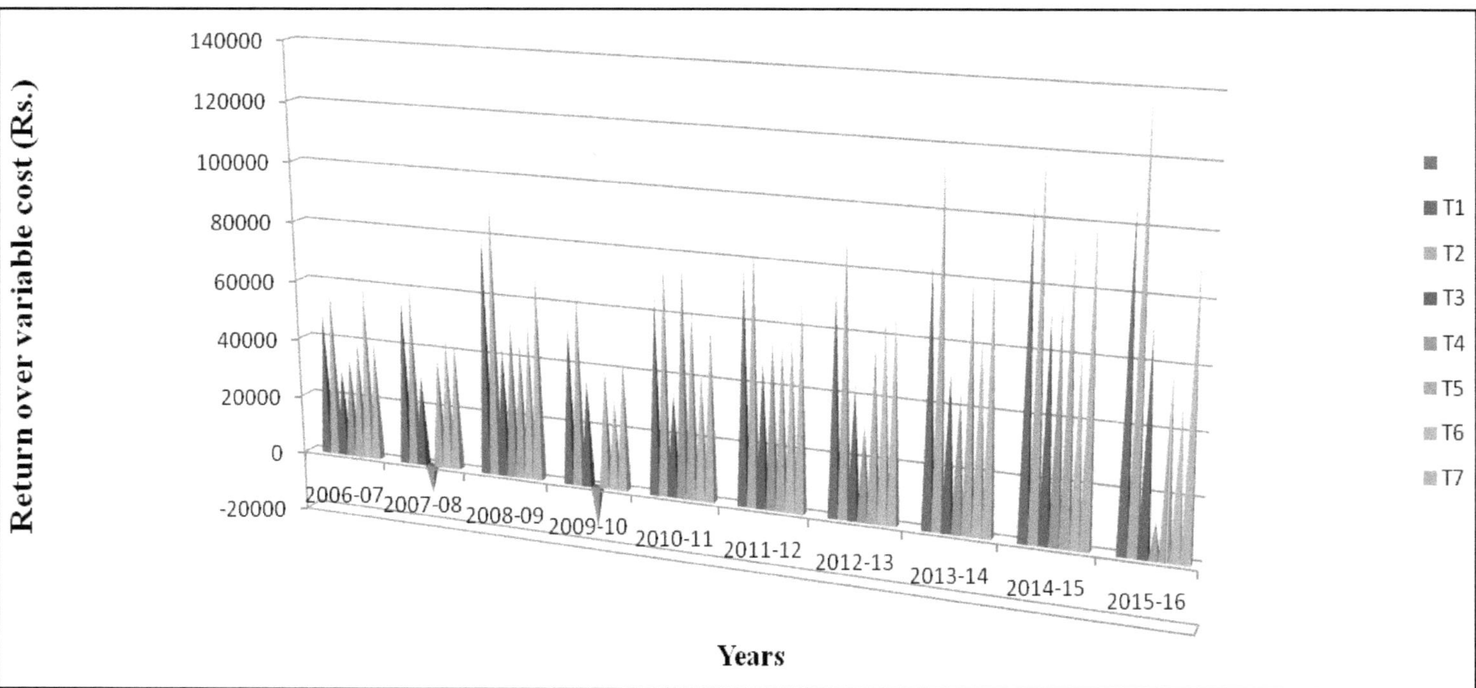

Figure 4: Return Over Variable Cost/Net Returns (Rs/ha) in different Cropping Systems for the Period of 2006-07 to 2015-16.

during 2007-08, 2008-09, 2010-11, 2011-12, 2012-13, 2013-14, 2014-15, 2015-16 and mean, respectively. During the year 2006-07 the maximum return over variable cost/ net returns of Rs. 58737 was recorded in pearl millet–wheat *desi*-cowpea cropping system.

Employment Generation

The data in Table 8 reveals that cotton engaged highest number of man days *i.e.* 116, 124, 131, 123, 138, 129, 135, 130, 125, 128 and 128 during 2006-07, 2007-08, 2008-09, 2010-11, 2011-12, 2012-13, 2013-14, 2014-15, 2015-16 and mean, respectively in *Kharif* season. Among *Rabi* crops, broccoli engaged higher man-days of 98, 101, 109, 105, 102, 106, 102, 104 and 95 during 2006-07, 2007-08, 2008-09, 2011-12, 2012-13, 2013-14, 2014-15 and mean, respectively (Table 9). Broccoli being vegetable crop engaged more man-days. Among summer crops the maximum man days of 69, 72, 68, 71, 104, 79, 82, 85, 88, 75 and 79 was engaged in onion during 2006-07, 2007-08, 2008-09, 2010-11, 2011-12, 2012-13, 2013-14, 2014-15, 2015-16 and mean, respectively (Table 10). Cluster bean-broccoli-onion engaged highest man days of 258 among all cropping systems and this cropping system keeps the farmers engaged for maximum time during the year.

Nutrients Uptake

During *Kharif* season, among the various crops, pearl millet removed maximum nitrogen (108.3 kg ha^{-1}) in pearl millet-wheat (*desi*)-cowpea cropping system followed by 106.6 kg nitrogen ha^{-1} by pearl millet in pearl millet-barley-mungbean cropping system (Table 11). Highest removal of phosphorus (27.3 kg ha^{-1}) was recorded by pearl millet in pearl millet–wheat (*desi*)–cowpea cropping system. The respective phosphorus removal by cluster bean and mungbean was 9.3 and 17.1 kg ha^{-1}. The potassium uptake during *Kharif* varied between 69.6 and 245.1 kg ha^{-1} in different crops, highest being by cotton in cotton-wheat cropping system.

During *Rabi* season the uptake of nitrogen, phosphorus and potassium in wheat ranged from 79.4 to 136.6, 13.5 to 29.1 and 136.7 to 170.4 kg ha^{-1}, respectively in different cropping system. Wheat (*desi*) removed 79.4, 13.5 and 136.7 kg ha^{-1} N, P and K, respectively. Barley removed 76.8 kg nitrogen ha^{-1}; 15.5 kg ha^{-1} phosphorus and 126.1 kg ha^{-1} potassium.

The total nutrients uptake (N+P+K) was highest in cluster bean-broccoli-onion and the value was 983.3 kg ha^{-1}. Pearl millet–wheat and cotton-wheat cropping systems recorded total nutrients (N+P+K) uptake of 665.8 and 761.5 kg ha^{-1}, respectively. Lowest total nutrient uptake was in mungbean-mustard+kashni (500.6 kg ha^{-1})

Soil Fertility Status

After the harvest of *Kharif*, *Rabi* and summer crops, soil samples were collected from each plot and were analyzed for pH, EC, OC and available nutrients status.

Table 8: Employment Generation (Man-days) during *Kharif* Season in different Cropping Systems for the Period of 2006-07 to 2015-16

Treat-ment	*Cropping systems*	*Man-Days Employment during Kharif Season*										
		2006-07	*2007-08*	*2008-09*	*2009-10*	*2010-11*	*2011-12*	*2012-13*	*2013-14*	*2014-15*	*2015-16*	*Mean*
T_1	Pearl millet – wheat	82	87	94	89	89	84	86	84	86	78	85.9
T_2	Cotton – wheat	116	124	131	123	138	129	135	130	125	128	127.9
T_3	Pearl millet – barley – mungbean	82	84	94	89	89	84	88	84	86	79	85.9
T_4	Clusterbean – broccoli – onion	74	77	81	79	94	88	92	90	88	72	83.5
T_5	Mungbean – mustard + kashni	71	74	69	61	66	61	65	66	65	58	65.6
T_6	Pearl millet – wheat (*Desi*) – cowpea	82	87	94	89	89	84	86	84	86	86	86.7
T_7	Pearl millet + mungbean – wheat + mustard	84	93	87	92	89	87	90	90	87	87	88.6

Table 9: Employment Generation (Man-days) during *Rabi* Season in different Cropping Systems for the Period of 2006-07 to 2015-16

Treat-ment	*Cropping Systems*	*Man-Days Employment during Rabi Season*										
		2006-07	*2007-08*	*2008-09*	*2009-10*	*2010-11*	*2011-12*	*2012-13*	*2013-14*	*2014-15*	*2015-16*	*Mean*
T_1	Pearl millet – wheat	85	90	93	91	91	92	98	98	98	93	92.9
T_2	Cotton – wheat	85	90	93	91	91	92	98	98	98	94	93.0
T_3	Pearl millet – barley – mungbean	72	75	68	64	72	69	72	72	75	68	70.7
T_4	Clusterbean – broccoli – onion	98	101	109	105	29	102	106	102	104	92	94.8
T_5	Mungbean – mustard + kashni	70	76	72	69	74	72	75	78	75	60	72.1
T_6	Pearl millet – wheat (*Desi*) – cowpea	81	86	69	67	72	68	72	70	72	65	72.2
T_7	Pearl millet + mungbean – wheat + mustard	85	91	93	95	91	95	98	101	92	60	90.1

Table 10: Employment Generation (Man-days) during Summer Season in different Cropping Systems for the Period of 2006-07 to 2015-16

Treat-ment	*Cropping Systems*	*Man-days Employment during Summer Season*										
		2006-07	*2007-08*	*2008-09*	*2009-10*	*2010-11*	*2011-12*	*2012-13*	*2013-14*	*2014-15*	*2015-16*	*Mean*
T_1	Pearl millet – wheat	-	-	-	-	-	-	-	-	-	-	-
T_2	Cotton – wheat	-	-	-	-	-	-	-	-	-	-	-
T_3	Pearl millet – barley – mungbean	41	44	39	47	37	33	35	38	32	18	36.4
T_4	Clusterbean – broccoli – onion	69	72	68	71	104	79	82	85	88	75	79.3
T_5	Mungbean – mustard + kashni	-	-	-	-	-	-	-	-	-	-	-
T_6	Pearl millet – wheat (*Desi*) – cowpea	51	54	47	46	45	38	43	40	42	35	44.1
T_7	Pearl millet + mungbean – wheat + mustard	-	-	-	-	-	-	-	-	-	-	-

Table 11: Mean (2006-07 to 2015-16) Nutrients Uptake (kg ha^{-1}) in different Cropping Systems at Hisar

Treatment		*N*	*P*	*K*	*Total*
T_1	Pearl millet	98.6	26.2	213.3	338.1
	Wheat	132.3	25.0	170.4	327.7
	Total	230.9	51.2	383.7	665.8
T_2	Cotton	151.0	32.9	245.1	429.0
	Wheat	136.6	29.1	166.8	332.5
	Total	287.6	62.0	411.9	761.5
T_3	Pearl millet	106.6	26.8	227.4	360.8
	Barley	76.8	15.5	126.1	218.4
	Mungbean	45.8	7.1	66.8	119.7
	Total	229.2	49.4	420.3	698.9
T_4	Cluster bean	59.5	9.3	69.6	138.4
	Broccoli	196.9	26.1	306.1	529.1
	Onion	141.2	32.9	141.7	315.8
	Total	397.6	68.3	517.4	983.3
T_5	Mungbean	105.7	17.1	100.1	222.9
	Mustard+*Kashni*	114.9	28.7	134.1	277.7
	Total	220.6	45.8	234.2	500.6
T_6	Pearl millet	108.3	27.3	223.9	359.5
	Wheat (*desi*)	79.4	13.5	136.7	229.6
	Cowpea	95.7	16.6	101.4	213.7
	Total	283.4	57.4	462.0	802.8
T_7	Pearl millet + Mungbean	95.9	23.1	188.3	307.3
	Wheat + Mustard	126.1	23.1	164.6	313.8
	Total	22.0	46.2	352.9	621.1

After the harvest of *Kharif* crops, the perusal of data in Table 12 revealed that in different treatments, pH and EC varied between 7.94 and 8.02 and between 0.17 and 0.19 dS $^{-1}$, respectively. The per cent organic carbon status ranged between 0.46 and 0.48. The available nitrogen level in different treatments ranged between 144.8 and 158.0 kg ha^{-}. The highest nitrogen of 158.0 kg ha^{-1} in soil was left in mungbean-mustard cropping system and minimum (144.8 kg ha^{-1}) in pearl millet–wheat (desi)-cowpea cropping system. The available phosphorus and potassium in different treatments ranged between 12.8 and 14.8 kg ha^{-1} and between 257.6 and 288.2 kg ha^{-1}, respectively. The available phosphorus was highest (14.8 kg ha^{-1}) in cluster bean-broccali-onion cropping system. Available potassium was lowest in 257.6 kg ha^{-1} in pearl millet–wheat. Highest potassium content of 288.2 kg ha^{-1} was again recorded in cluster bean – broccoli - onion cropping system.

Table 12: Mean (2006-07 to 2015-16) Effect of Cropping Systems on Soil Fertility Status after *Kharif* Crop

Treatment	pH	E.C. (dS m⁻¹)	O.C. (Per cent)	Available Nutrients (kg ha⁻¹)		
				N	P	K
T_1	8.02	0.18	0.46	147.4	13.1	257.6
T_2	7.96	0.19	0.46	146.7	13.9	279.2
T_3	8.01	0.19	0.47	146.3	12.8	272.6
T_4	7.94	0.17	0.48	152.1	14.8	288.2
T_5	8.01	0.18	0.48	158.0	14.3	282.4
T_6	8.02	0.19	0.46	144.8	13.3	270.2
T_7	7.96	0.18	0.46	151.8	13.4	273.8

Table 13: Mean (2006-07 to 2015-16) Effect of Cropping Systems on Soil Fertility Status after *Rabi* Crop

Treatment	pH	E.C. (dS m⁻¹)	O.C. (Per cent)	Available Nutrients (kg ha⁻¹)		
				N	P	K
T_1	7.73	0.21	0.47	164.6	13.0	283.7
T_2	7.76	0.20	0.46	163.7	13.2	287.1
T_3	7.77	0.20	0.47	169.8	13.2	277.9
T_4	7.81	0.20	0.48	161.4	13.2	277.9
T_5	7.80	0.20	0.48	159.4	12.5	288.8
T_6	7.78	0.21	0.48	172.6	13.5	277.8
T_7	7.84	0.20	0.47	166.5	13.1	283.9

Table 14: Mean (2006-07 to 2015-16) Effect of Cropping Systems on Soil Fertility Status after Summer Crops Harvest

Treatment	pH	E.C. (dS m⁻¹)	O.C. (Per cent)	Available Nutrients (kg ha⁻¹)		
				N	P	K
T_1	-	-	-	-	-	-
T_2	-	-	-	-	-	-
T_3	7.83	0.20	0.47	164.3	12.6	263.4
T_4	7.79	0.21	0.48	162.5	13.6	268.8
T_5	-	-	-	-	-	-
T_6	7.78	0.21	0.47	167.6	13.5	279.4
T_7	-	-	-	-	-	-

After the harvest of *Rabi* crops, a cursory look at the data presented in Table 13 revealed that pH, EC and organic carbon (per cent) among different treatments ranged between 7.73 and 7.84, 0.20 and 0.21 dS m^{-1} and 0.46 and 0.48, respectively. Available nitrogen status among different cropping systems ranged from 159.4 to 172.6 kg ha^{-1} being highest in pearl millet–wheat (desi)-cowpea cropping system and lowest in mungbean-mustard cropping system. The available phosphorus among different treatments ranged narrowly from 12.5 to 13.5 kg ha^{-1} while potassium content ranged between 277.8 to 288.8 kg ha^{-1} among different cropping systems. The potassium content was comparatively higher in mungbean–mustard + kashni and cotton–wheat cropping systems.

After harvest of the summer crop pH, EC and organic carbon (per cent) among different treatments (Table 14) ranged between 7.78 to 7.83; 0.20 to 0.21 and 0.47 to 0.48, respectively. The available nitrogen and potassium status after harvest of summer crop was highest in pearl millet–wheat (desi)-cowpea cropping system, while phosphorus was highest in cluster bean–broccoli-onion cropping system.

Dynamics of Weed Flora

The weed flora in different cropping systems was studied and details for *Kharif* and *Rabi* weeds is given in Tables 15 and 16, respectively.

Table 15. Common Weeds during *Kharif* under different Cropping Systems

Sl.No.	*Cropping System*	*Name of the Common Weed(s)*
1.	Pearl millet-wheat	Itcit, Motha, Doob, Kondhra, Cholai, Sawank
2.	Cotton – wheat	Motha, Doob, Santhi, Bhakhri, Kundra, Sawank, Chilmil
3.	Pearl millet – barley – mung bean	Sawank, Santhi, Kundra, Motha, Bhakhri, Doob
4.	Cluster bean – broccoli – onion	Kundra, Santhi, Doob, Nooni, Bhakhri, Motha
5.	Mung bean – mustard + kashni	Santhi, Kundra, Doob, Sawank, Bhakhi, Cholai
6.	Pearl millet – wheat (*desi*) – cowpea (Veg.+R)	Kundra, Motha, Doob, Bhakhri, Cholai, Nooni, Sawank
7.	Pearl millet + mung bean – Wheat + mustard	Kundra, Motha, Santhi, Bhakhri, Cholai, Sawank, Doob

Table 16: Common Weeds during *Rabi* Season under different Cropping Systems

Sl.No.	*Cropping System*	*Name of the Common Weed(s)*
1.	Pearl millet-wheat	Bathu, Krishan Neel, Hiran Khuri, Pithpapra, Janglipalak, Maina, Kandai, Jangli Jai, Kanki, Chatri, Matri, Metha, Senji, Jangli gobhi, Jangli dhainia, Gajri.
2.	Cotton–wheat	Krishan Neel, Bathu, Metha, Matri, Maina, Pitpapra, Jangli Palak, Hiran Khuri, Chatri, Gajrela, Kandai
3.	Pearl millet –barley–mung bean	Bathu, Krishan Neel, Metha, Matri, Chatri, Hiran Khuri, Jangli Jai, Pitpapra, Maina
4.	Cluster bean–broccoli–onion	Bathu, Krishan Neel, Metha, Maine, Jangli Palak, Chatri, Doob, Senji
5.	Mungbean–mustard+kashni	Bathu, Krishan Neel, Maina, Gajrela, Kandai, Jangali Palak, Metha, Matri, Chatri
6.	Pearl millet–wheat(*desi*)–cowpea (Veg.+R)	Chatri, Matri, Bathu, Krishan Neel, Motha, Senji, Gajrela, Doob, Matri, Kandai, Maina, Pitpapra, Jangli Jai
7.	Pearl millet+mung bean–wheat+mustard	Bathu, Kanki, Metha, Pitpapra, Krishan Neel, Gajri, Hiran Khuri, Jangli Jai, Jangli Palak, Matri, Chatri, Kandai, Maine, Doob, Senji

Acknowledgement

The authors feel highly indebted to the Indian Institute of Farming Systems Research, ICAR, Modipuram for providing technical guidance and financial support to carry out this experiment for such a long time in the Department of Agronomy, CCS Haryana Agricultural University, Hisar. The authors also thankfully acknowledge the scientific staff worked in the project from 2006-07 till date for their contribution in the conduct of this experiment.

Chapter 6

Nutrient Management

Long Term Effect of Integrated Nutrient Management on Yield Sustainability, Economic Viability and Soil Health under Pearl millet-Wheat Cropping System in Haryana

Food security for a huge country like India, with high density of population in general and below poverty line is of vital importance. The population in India with a growth rate of about 2.3 percent has crossed one billion at the beginning of the century. India has to produce around 300 mt of food grains by 2025 A.D. to nourish over 1.4 billion populations from 0.15 hactare land per capita or less. To feed this ever-increasing huge population, many intensively cropped cereal based cropping systems are under cultivation in the country.

Last four decades have witnessed a phenomenal growth in production and productivity of Indian agriculture. Today, India has sufficient food to feed three times as many people as at the time of independence. The food grain production quadrupled from a low of 51.0 million tones (mt) in 1950-51 to 259.6 mt in 2012-13 (Anonymous, 2013). Food security has improved for consumers who are at lower strata of income, employment opportunities increased for the landless agricultural labourers and indirectly development of small scale rural industries received a boost.

Pearl millet [*Pennisetum glaucum* (L.) R. Br. emend. Stuntz.]-wheat (*Triticum aestivum* L.) is one of the important cropping systems of the country and spreads over (i) arid eco-region comprising, western plains, Kachh and parts of Kathiawar Peninsula having desert and saline soils representing Gujarat, Rajasthan and Haryana; (ii) semi-arid eco-region comprising northern plains of Haryana, western Uttar Pradesh and central high lands of Rajasthan with alluvium derived soils. This system is very exhaustive and a crop giving 2.9 tones per hactare of pearl millet and 4.2 tones per hactare of wheat which may remove 238, 54 and 131 kg nitrogen, phosphorus and potassium per hactare, respectively (Hegde *et al.*, 1992).

During the last five decades, consumption of chemical fertilizers (NPK) has increased phenomenally from 0.07 mt in 1950-51 to 24.91 mt in 2008-09 (Anonymous, 2009). Consumption of fertilizers (all nutrients) per hactare increased from 1.0 kg to 106 kg (Hegde and Babu, 2004). There is a considerable gap between production and consumption of chemical fertilizers in the country. Therefore, currently Indian agriculture is operating on a negative balance of plant nutrients at the rate of 8-10 mt per annum. With the application of recommended dose of fertilizers yield potential of this cropping system (cereal-cereal) has reached to a plateau because of deteriorated soil health and especially organic matter depletion.

Long term studies being carried out at several locations on different cropping systems indicated that application of all the needed nutrients through chemical fertilizers has deleterious effect on soil health, leading to unsustainable yields (Hegde and Katyal, 1998; Behra *et al.*, 2007). This further has led to aggravated micro-nutrient deficiency in soil system.

Since, the nutrient turnover in soil-plant system is considerably high under intensive cropping system. So, neither the chemical fertilizers nor the organic/ biological sources alone can achieve production sustainability. Even with the so called balance use of NPK fertilizers in long term studies, higher yield levels could not be maintained for years because of emergence of secondary and micro-nutrient deficiency and deterioration in the soil physical environment. Whereas, organic manure alone or in combination with inorganic fertilizers is known to have favourable effect on soil environment and correct marginal deficiency of secondary and micro-nutrients and enhance efficiency of applied nutrients. Therefore, there is need to improve nutrient supply system for sustainable production of this very important cropping system of India.

For higher fertilizer use efficiency and sustainability of cropping system, there is need to recommend and develop site specific nutrient management strategies considering the cropping system as a whole, instead of component crops in isolation (Khurana and Singh, 2008 and Singh *et al.*, 2008). To achieve this, we have to take into account the direct as well as residual effect of fertilizer to different crops in the system.

A lot of information is available on response of pearl millet and wheat to fertilizer application as sole crop but as a cropping system limited information is available. It is being realized that system based research would be more advantageous for optimizing the use of different sources of plant nutrients. Therefore, keeping in view the above facts present investigation "Response of pearl millet-wheat cropping system to various sources of nutrients in terms of growth, yield and nutrient uptake" was carried out at Research Farm of Department of Agronomy, CCS HAU, Hisar since 1984-85 with the objectives; "To develop suitable integrated nutrient supply and management system for cereal based cropping system with judicious combination of organic manures on the productivity of cereal based crop sequence and on soil health".

Methodology

Experimental Site and Location

The experimental site at Research Farm of Chaudhary Charan Singh Haryana Agricultural University, Hisar is situated in semi-arid, sub-tropics at 29.10°N latitude and 75.46°E longitude at an altitude of 215.2 meters above mean sea level, in Haryana state of India.

Treatments under the Experiment

The field experiment was laid out in randomized block design with four replications at Agronomy Research Farm of CCS Haryana Agricultural University, Hisar. The treatment details are presented in Table 1.

Table 1: Details of Integrated Nutrient Management Treatments in Pearl Millet-Wheat Cropping System

Treat-ment	*Season*	
	Kharif	*Rabi*
T_1	Control (no fertilizer)	Control (no fertilizer)
T_2	50 per cent recommended NPK dose through fertilizers	50 per cent rec. NPK dose through fertilizers
T_3	50 per cent recommended NPK dose through fertilizers	100 per cent NPK dose through fertilizers
T_4	75 per cent recommended NPK dose through fertilizers	75 per cent rec. NPK dose through fertilizers
T_5	100 per cent recommended NPK dose through fertilizers	100 per cent rec. NPK dose through fertilizers
T_6	50 per cent recommended NPK dose through fertilizers + 50 per cent N through FYM	100 per cent rec. NPK dose through fertilizers
T_7	75 per cent recommended NPK dose through fertilizers + 25 per cent N through FYM	75 per cent rec. NPK dose through fertilizers
T_8	50 per cent recommended NPK dose through fertilizers + 50 per cent N through wheat straw	100 per cent rec. NPK dose through fertilizers
T_9	75 per cent recommended NPK dose through fertilizers + 25 per cent N through wheat straw	75 per cent rec. NPK dose through fertilizers
T_{10}	50 per cent recommended NPK dose through fertilizers + 50 per cent N through G.M.	100 per cent rec. NPK dose through fertilizers
T_{11}	75 per cent recommended NPK dose through fertilizers + 25 per cent N through G.M.	75 per cent rec. NPK dose through fertilizers
T_{12}	Farmer's Practice	Farmer's Practice

Overview of Different INM Treatments.

Overview of Different INM Treatments.

Overview of Different INM Treatments.

Overview of Different INM Treatments.

Treatment Details

Design: Randomized block design

Replications: 4

Total treatments: 12

Total number of plots: 48

Gross plot size: 10 x 8 m^2

Net plot size: Pearl millet : 8.2 x 6 m^2

Wheat : 8.4 x 6 m^2

Methods followed for Soil and Plant Analysis

Sl.No.	*Character under Analysis*	*Method Used*	*Reference*
A. Soil analysis			
1.	Available nitrogen (kg ha^{-1})	Alkaline permanganate method	Subbiah and Asija (1956)
2.	Available phosphorus (kg ha^{-1})	Olsen's method	Olsen *et al.* (1954)
3.	Available potassium (kg ha^{-1})	Flame photometer method	Jackson (1973)

Sl.No.	Character under Analysis	Method Used	Reference
B. Plant analysis			
1.	Nitrogen (per cent)	Kjeldahl method	
2.	Phosphorus (per cent)	Ammonium-vanadomolybdo phosphoric acid yellow colour method	Jackson (1973)
3.	Potassium (per cent)	Flame emission spectrometric method	Jackson (1973)

Economic Studies

Cost of cultivation and gross returns (Rs ha^{-1}) for different treatments were calculated on the basis of the approved market rates for inputs and rates of output/produce fixed by Director of Farms, Chaudhary Charan Singh Haryana Agricultural University, Hisar. Returns over variable cost/net returns (Rs ha^{-1}) were worked out by subtracting the total cost of cultivation of each treatment from the gross income of respective treatment. B : C was also worked out to ascertain the economic viability of different treatments.

Statistical Analysis

All the experimental data reported are the mean values of different parameters measured. The statistical method described by Panse and Sukhatme (1985) was followed for statistical analysis and interpretation of the experimental results. In order to evaluate the comparative performance of different treatments, the data was analyzed by the technique of analysis of variance given by Fisher (1958). To judge the significance of the difference between two treatments, critical difference (C.D.) was worked out by the following formula:

$$C.D. = \sqrt{\frac{2 \times \text{Mean square error}}{n}} \times \text{'t'}$$

where,

C.D. = Critical difference.

n = Number of replications of the factor for which C.D. is to be calculated.

t = The value from fisher table for error degree of freedom at 5 percent level of significance.

Results and Discussion

Crop Productivity

Pearl millet

During 1984-85 to 88-89 pearl millet yield was higher (2436 kg/ha) with the application of 100 per cent recommended dose of NPK. The FYM application to substitute 25 per cent N + recommended NPK in pearl millet and 75 per cent recommended NPK in wheat yielded better over all other sources and their rate of application of organic source of nutrients. FYM application for 25 per cent N requirement gave higher yield of 3.87, 20.82, 22.10, 15.28 and 2.73 per cent respectively over 50 per cent N from FYM, 50 per cent N through wheat straw, 25 per cent N through wheat straw, 50 per cent N through green manure and 25 per cent N through green manure, respectively. 100 per cent recommended NPK produced 6.77 per cent higher yield over the treatment where 25 per cent N was replaced with FYM and 75 per cent NPK was applied through chemical fertilizer in pearl millet and 75 per cent NPK was applied in wheat. During this period, 53.28 per cent low yield was produced over 100 per cent recommended NPK to both the crops. Katyal *et al.* (2002) has also reported that with the application of 100 percent NPK through chemical fertilizers to both pearl millet and wheat gave significantly higher yield of pearlmilet (2.38 t/ha) and wheat (4.08 t/ha). During 1989-90 to 1993-94 almost same trend was observed in terms of grain yield of pearl millet. After 10 years *i.e.* from 1994-95 to 1998-99 the effect of green manure and wheat straw were recorded and the grain yield increase to the tune of 13.345 and 6.83 per cent with the application of 25 per cent N through wheat straw and green manure respectively, over substitution of 25 per cent N through FYM (Table 2).

During 1999-2000 to 2003-04, the grain yield of pearl millet was 2600 kg/ha and it was 6.3, 14.11 and 20.88 per cent higher over the yield of last three years, with 100 per cent chemical fertilizers in both crops. From here *i.e.* after about 15 years of experimentation, the increase in yield was 401, 554 and 752 kg/ha with fertilization of 50 per cent N through FYM in pearl millet where grain yield was recorded 2584 kg/ha. At this stage of experimentation, the yields were almost same (2584 and 2586 kg/ha) with the substitution of N through FYM to the tune of 50 per cent and 25 per cent. Earlier, 25 per cent N substitution was giving high yield over 50 per cent N through FYM. At this stage the application 50 per cent N through green manure has taken edge over 25 per cent N with green manure, which shows that larger substitution (50 per cent) through FYM and green manure has started accumulation organic matter in soil. The increase in yield after 15 years from control to 100 per cent has been recorded 168 per cent, which was 114, 109 and 110 per cent higher in 1984-85 to 88-89, 1989-90 to 1993-94 and 1994-95 to 98-99, respectively, indicating the deficiency of nutrients for crop because of no fertilizer application. During 2004-05 to 2008-09 and 2009-10 to 2013-14 the yield was recorded 4.51 and 4.34 per cent higher where 50 per cent N was substituted through FYM along with 50 per cent NPK through chemical fertilizers in pearl millet and 100 per cent NPK to both the crops. Similar pattern was observed during 2014-15 to 2015-16. Hegde (1998) has also reported that 50 per cent N needs of pearl millet can be met from FYM. At the start of experiment *i.e.* 1984-85 to 1988-89 the increase in pearl millet yield with increase in fertilizer dose *i.e.* from 50 to 75 per cent and 75 to 100 per cent was 9.95 and 20.53 per cent, which was 15.43 and 4.34 per cent at end of the experiment.

Wheat

During 1984-85 to 88-89 the wheat yield was recorded highest (4110 kg/ha) with 50 per cent recommended NPK through chemical fertilizers + 50 per cent N through green manure in pearl millet followed by 100 per cent recommended NPK through chemical fertilizers in wheat and it was followed by 100 per cent recommended

Table 2: Effect of INM Treatments on Pearl Millet and Wheat Yield in Pearl Millet-Wheat Cropping System (Means of 5-years)

Treatment	*Pearl Millet (kg ha^{-1})*							*Wheat (kg ha^{-1})*						
	84-85 to 88-89	*89-90 to 93-94*	*94-95 to 98-99*	*99-00 to 03-04*	*04-05 to 08-09*	*09-10 to 13-14*	*14-15 to 15-16*	*84-85 to 88-89*	*89-90 to 93-94*	*94-95 to 98-99*	*99-00 to 03-04*	*04-05 to 08-09*	*09-10 to 13-14*	*14-15 to 15-16*
T_1	1138	1067	979	969	1090	966	953.5	1586	1225	1549	1065	1229	1218	1038
T_2	1727	1529	1573	1745	1979	2088	1894	3123	2722	3320	3172	3565	3705	3735
T_3	1838	1576	1608	1858	2092	2216	2093	4037	3522	3930	4497	4775	5241	5038
T_4	2021	1762	1795	2174	2510	2710	2548	3771	3298	3689	4140	4401	4615	4792
T_5	2436	2233	2057	2600	3005	3133	3059	3983	4081	4171	4872	5101	5527	5651
T_6	2183	2030	1832	2584	3147	3288	3231	4200	3834	4231	4684	5233	5682	5764
T_7	2271	2183	1903	2586	2941	3053	2855	3815	3564	3737	4505	4681	4800	4992
T_8	1798	1580	1663	2303	2487	2577	2759	3945	3442	3906	4360	4888	5329	5467
T_9	1769	1925	2158	2421	2675	2773	2494	3727	3397	3806	4248	4616	4665	4769
T_{10}	1924	2040	1820	2549	2787	3062	2979.5	4110	3899	4061	4196	5017	5564	5641
T_{11}	2209	2102	2033	2492	2885	3059	2837	3667	3635	3814	4362	4723	4712	4768
T_{12}	1674	1667	1591	1601	2353	2821	2742.5	2781	2944	3520	3800	4598	5132	5232

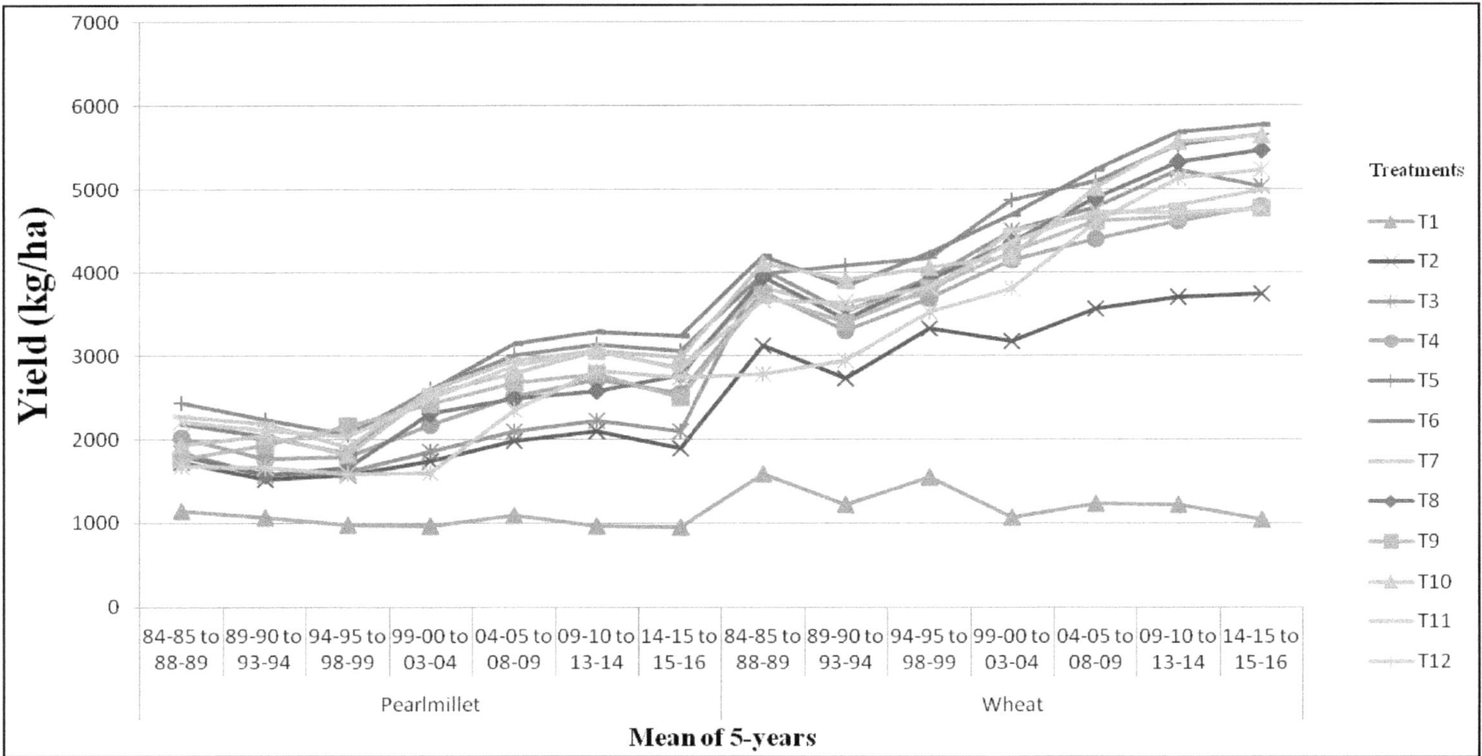

Figure 1: Mean Pearl Millet and Wheat Yield in Pearl Millet-Wheat Cropping System from 1984-85 to 2015-16.

NPK (3983 kg/ha) in both the plots (Table 2). After 5 years of experimentation *i.e.* from 1989-90 to 93-94, highest grain yield (4081 kg/ha) was recorded with the application of 100 per cent recommended NPK in both the crops. Among organic sources of nutrients, green manuring along with chemical fertilizers followed by 100 per cent recommended NPK through chemical fertilizers yielded more as to FYM and wheat straw. During 1994-94 to 1998-99 substitution of 50 per cent N through FYM yielded (4231 kg/ha) higher over all the organic sources of nutrients applied with chemical fertilizers and 100 per cent recommended NPK to both the crops. It was 1.41, 7.68 and 4.02 per cent higher over 100 per cent recommended NPK through chemical fertilizers, 50 per cent N substitution with wheat straw and green manure, respectively. The wheat grain yield was recorded 132 and 146 kg/ha higher where 50 per cent N was substituted with FYM in pearl millet over 100 per cent recommended NPK through chemical fertilizers in both the crops during 2004-05 to 08-09 and 2009-10 to 2010-11, respectively. Whereas, the wheat yield was higher with the application of 100 per cent recommended NPK during 1999-2000 to 2003-04. Over the years, slightly higher wheat grain yield was recorded with the application of 50 per cent N through FYM + 50 per cent recommended NPK through chemical fertilizers in pearl millet, followed by 100 per cent recommended NPK in wheat. Patil *et al.* (1995) has also reported that 50 per cent N substitution through FYM had significant residual effect and thereby, the productivity of wheat in pearl millet – wheat cropping system increased. Among FYM, wheat straw and green manure application to substitute 50 per cent N in pearl millet along with chemical fertilizers, FYM performed better followed by green manure. The increase in yield was very high from 75 per cent to 100 per cent recommended NPK in both the crops and it was 212, 783, 482, 732, 700 and 894 kg/ha during 1984-85 to 88-89, 1989-90 to 93-94, 1994-95 to 98-99, 1999-00 to 03-04, 2004-05 to 08-09 and 2009-10 to 13-14, respectively. Similar pattern was also observed during 2014-15 to 2015-16. Hegde and Katyal (1999) on the basis of long term experiment has also reported that both pearl millet and wheat responded significantly to addition of N upto 120 kg/ha. The grain yield in control (un-fertilized) decreased from start of experiments to end of experiments which resulted in nutrient deficiency showing that fertilizer application is necessary for harvesting good yield.

Wheat Equivalent Yield (WEY) kg/ha

The WEY from 2002-03 to 2012-13 was higher where 50 per cent N was substituted through FYM + 50 per cent recommended NPK through chemical fertilizers in pearl millet followed by 100 per cent recommended NPK though chemical fertilizers in wheat. It has also been reported by Hegde (1998) that in total productivity of the system, only FYM could replace 50 per cent N needs of pearl millet with out much adverse effect on production. Such type of results has also been reported by Katyal *et. al* (1999). They further stated that this may be attributed to the addition of secondary and micronutrients and prolonged release of N from organic sources which might have resulted better and ultimately higher system productivity. The increase in WEY over 100 per cent recommended NPK in both crops was 4.38, 2.29, 3.08, 3.57 and 2.94 per cent during 2002-03 to 2003-04, 2004-05 to 2005-06, 2006-07 to 2007-08, 2008-09 to 2009-10,2010-11 to 2011-12 and 2012-13, respectively. This indicates that FYM application along with chemical fertilizer have beneficial effect on productivity of pearl millet- wheat cropping system. WEY have also followed the same trend as that of wheat grain yield with the application of 50 per cent N through FYM, wheat straw and green manure, being highest with FYM, followed by green manure and lowest yield with wheat straw. With every 25 per cent incremental N application *i.e.* from 50 per cent - 75 per cent the increase in yield was 33.98, 31.66, 23.46, 21.77 and 26.97 per cent ; and from 75 per cent - 100 per cent recommended NPK the increase in yield was 11.23, 13.70, 18.59, 21.23 and 17.35 per cent during 2002-03 to 2003-04, 2004-05 to 2005-06, 2006-07 to 2007-08, 2008-09 to 2009-10,2010-11 to 2011-12 and 2012-13, respectively showing that higher yield can be harvested with every kg nutrient application up to recommended dose of nutrients. Highest mean wheat equivalent yield to the tune of 9007 during 2013-14 to 2014-15 and 8241 kg/ha during 2015-16 was recorded in treatment T_6. Both the crops in the system responded to increased fertilizer level upto 120 kg/ha as has been reported by Hegde and Katyal (1999)

Return over Variable Cost/Net Return (Rs/ha)

Starting from 2002-03 to 2003-04 and towards the end of experimentation *i.e.* 2010-11 to 2012-13, the return over variable cost/net returns were higher every year with 50 per cent N application through FYM in pearl millet except during 2002-03 to 2003-04. The increase in return over variable cost/net returns with the application of each incremental nitrogen *i.e.* from 50 per cent -75 per cent were 78.40, 57.02, 55.86, 67.39 and 53.34 per cent and from and 75 per cent - 100 per cent were 27.23, 20.44, 29.0, 35.47 and 24.22 per cent, respectively. 50 per cent N substitution from FYM in pearl millet gave 608, 3287, 5205 and 1289 Rs/ha over 100 per cent chemical fertilizer application in both the crops during all the years except during 2002-03 to 2003-04, where 100 per cent chemical fertilizer in both the crops gave higher return over variable cost/net returns of Rs.697/ha. From 2002- 03 to 2003-04, the return over variable cost/net returns increased every year irrespective of source of nutrient applied to crops. This might be due to increase in yield and minimum support price of the crops. the per cent increase in return over variable cost/net returns from 25 per cent N substitution to 50 per cent N substitution through FYM were 11.06, 7.06, 14.49, 24.12 and 11.67 per cent during 2002-03 to 2003-04, 2004-05 to 2005-06, 2006-07 to 2007-08, 2008-09 to 2009-10, 2010-11 to 2011-12 and 2012-13, respectively. Whereas, return over variable cost/net returns were higher where 25 per cent N was substituted through wheat straw over 50 per cent N substitution and it was 6.51, 9.50, 27.78, 14.24. and 4.71 per cent. Green manuring followed the same trend as that of FYM application during all the years except 2004-05 to 2005-06 and percent increase in yield was2.69, 5.73,13.11 and 12.65 respectively. Highest return over variable cost/net return of Rs. 98767 during 2013-14 to 2014-15 and Rs. 101079 during 2015-16 was recorded in the treatment where 50 per cent recommended NPK dose through fertilizers + 50 per cent N through FYM during *Kharif* and 100 per cent rec. NPK dose through fertilizers during *Rabi* season. With no application of fertilizer, the pearl millet-wheat cropping system was not remunerative (Table 3).

Table 3: Effect of INM Treatments on Wheat Equivalent Yield and Return Over Variable Cost/ Net Returns in Pearl Millet-Wheat Cropping System (Means of 2-years)

Treatment	Wheat Equivalent Yield (kg ha^{-1})							Return Over Variable Cost/Net Returns (Rs ha^{-1})						
	02-03 to 03-04	04-05 to 05-06	06-07 to 07-08	08-09 to 09-10	10-11 to 12-13	13-14 to 14-15	15-16	02-03 to 03-04	04-05 to 05-06	06-07 to 07-08	08-09 to 09-10	10-11 to 12-13	13-14 to 14-15	15-16
T_1	1951	1768	1972	2073	2047	1959	1794	-3081	352	4737	-8054	372	-5800	-7462
T_2	4458	4191	4936	5833	5201	5727	5098	14568	16720	24807	28863	38162	46613	47390
T_3	5705	5329	6244	7530	6777	7501	6245	22619	24679	38185	51341	59546	73451	65941
T_4	5973	5518	6094	7103	6604	7300	6684	25989	26255	38665	48313	58518	73252	75025
T_5	6644	6274	7227	8611	7750	8643	8027	33067	31622	49874	65448	72689	91464	96999
T_6	6948	6421	7457	8930	7985	9007	8241	32370	32230	53161	70653	73978	98767	101079
T_7	6568	6075	6717	7729	7035	7704	7242	29145	30105	46434	56924	66245	80974	83175
T_8	6287	5824	6677	7826	7118	8006	7704	24455	23196	29659	36569	50591	65197	92336
T_9	6200	5752	6586	7291	6703	7273	6716	26047	25400	37894	41779	52974	56965	77482
T_{10}	6575	6047	6959	8452	7745	8654	7842	29787	28810	47416	65783	69513	95029	95192
T_{11}	6402	5938	6775	8029	6936	7667	6874	28985	29196	44699	57157	60718	78021	78903
T_{12}	5304	5193	6342	7748	7233	7821	7447	21384	22644	40148	55804	67930	79652	86106

Nutrients Uptake

Nitrogen Uptake

Nitrogen uptake by pearl millet (Table 4) in control reduced to half as we move from mean of 1984-85 to 1988-89 (40.7 kg ha^{-1}) to mean of 2009-10 to 2013-14 (22.2 kg ha^{-1}). Comparing means of T_2 and T_3 treatments in the beginning (84-85 to 88-89) with the means at the end (09-10 to 13-14) shows that there was a decline in nitrogen (N) uptake by pearl millet. Means of treatments T_6 and T_{10} showed marked increase in magnitude in the beginning and at the end. In rest of the treatments, marked variation with respect to above mentioned trend has not been observed. Comparison of per cent increase in N uptake by pearl millet in T_6 and T_{10} with T_1 during the mean of year 84-85 to year 88-89 showed an increase of 77.1 and 56.1. While, mean of 09-10 to 13-14 showed, respective, per cent increase in N uptake of 307.5 and 250.9 kg ha^{-1} in T_6 and T_{10} over T_1. Mean during 09-10 to 13-14 showed an increase of 4 and 4.5 times in T_6 and T_{10} over T_1 as compared to mean during the period from 84-85 to 88-89.

Nitrogen uptake by wheat in T_1 did not show any variation across the 5-years means of different years (Table 4). In rest of the treatments, marked increase has been observed. Across 5-years mean as we move from the mean of of 84-85 to 88-89 towards the mean of 09-10 to 13-14, the per cent increase in N uptake ranged between 42 in T_2 and 91 in T_6. Similar pattern was observed during 2014-15 to 2015-16. Similar results were observed by Oo *et al.*, 2007) which shows that the nitrogen uptake was lowest in control plot during both the seasons due to low yield of crops in this treatment and the nitrogen uptake in both pearl millet and wheat increased with the increase in fertilizer dose from 50 to 75 per cent and from 75 to 100 per cent of recommended dose.

Phosphorus Uptake

Mean phosphorus (P) uptake by pearl millet (Table 5) in control has reduced to less than one-third on comparison of mean uptake of 09-10 to 13-14 with the mean of 84-85 to 88-89. In all treatments mean phosphorus (P) uptake has reduced to almost one-half in the mean of 89-90 to 93-94 over the mean of 84-85 to 88-89. As we proceed forward towards the mean of 94-95 to 98-99 there was a further decline in P uptake by pearl millet. Thereafter, a plateau of increase in P uptake has been observed. Trend in treatments T_5, T_6, T_7 showed similar dominance in P uptake both in the mean P uptake during 84-89 to 88-89 as well as mean P uptake during 09-10 to 13-13. There was a decline in mean P uptake of 28.4, 16.2 and 35.7 per cent, in the mean of 09-10 to -14 over the mean of 84-85 to 88-89 in treatments T_5, T_6 and T_7, respectively. Whereas, decline in P uptake in treatments T_2 and T_3 in the mean of 09-10 to 13-14 over the mean of 84-85 to 88-89 was 41.4 and 34.3 per cent, respectively.

On the contrary, increase in P uptake by wheat (Table 5) in mean of 09-10 to 13-14 over the mean of 84-85 to 88-89 has been observed in all the treatments except in T_1 and T_2 where only 50 per cent recommended NPK dose through fertilizers both during *Kharif* and *Rabi* were applied. Among integrated nutrient management (INM) treatments, T_6 showed maximum increase of 54.4 per cent followed by T_{10}

Table 4: Effect of INM Treatments on Nitrogen Uptake by Pearl Millet and Wheat in Pearl Millet-Wheat Cropping System (Means of 5-years)

Treatment	*Nitrogen Uptake by Pearl Millet (kg/ha)*							*Nitrogen Uptake by Wheat (kg/ha)*						
	84-85 to 88-89	*89-90 to 93-94*	*94-95 to 98-99*	*99-00 to 03-04*	*04-05 to 08-09*	*09-10 to 13-14*	*14-15 to 15-16*	*84-85 to 88-89*	*89-90 to 93-94*	*94-95 to 98-99*	*99-00 to 03-04*	*04-05 to 08-09*	*09-10 to 13-14*	*14-15 to 15-16*
T_1	40.7	25.0	26.0	19.6	25.4	22.2	18.9	30.3	18.3	32.4	22.5	25.0	24.9	21.1
T_2	60.4	35.4	42.9	36.0	52.0	46.4	38.3	55.0	45.3	65.9	65.1	77.1	77.9	77.8
T_3	67.0	33.2	42.7	40.9	57.1	48.0	46.1	66.7	54.7	75.2	98.7	105.0	116.6	108.3
T_4	64.8	36.6	50.8	50.5	67.8	67.9	59.8	63.7	55.1	71.8	98.7	97.8	102.4	107.0
T_5	75.6	53.3	56.0	60.8	117.9	81.3	77.8	66.0	63.3	80.7	118.6	123.4	131.4	130.5
T_6	72.1	42.3	61.9	63.0	95.0	88.3	82.5	72.7	62.5	78.5	115.9	137.0	139.7	139.7
T_7	82.1	50.5	61.4	56.1	89.7	74.6	67.0	62.1	62.1	73.8	108.5	109.8	110.2	113.8
T_8	55.2	36.6	47.7	49.7	71.5	58.8	62.2	66.7	56.9	75.0	110.3	113.1	121.8	124.2
T_9	59.9	40.0	47.7	52.6	73.8	65.7	57.0	61.1	55.9	73.3	110.2	101.1	106.1	110.8
T_{10}	63.5	44.6	49.4	58.3	79.6	74.7	69.8	68.7	68.6	79.7	106.2	128.7	131.5	132.4
T_{11}	72.6	52.6	59.0	63.4	85.1	76.2	68.5	57.9	66.1	74.1	106.2	107.6	108.9	111.6
T_{12}	54.8	41.7	44.6	36.5	63.9	67.1	59.5	46.4	52.9	69.5	85.8	103.8	118.2	123.3

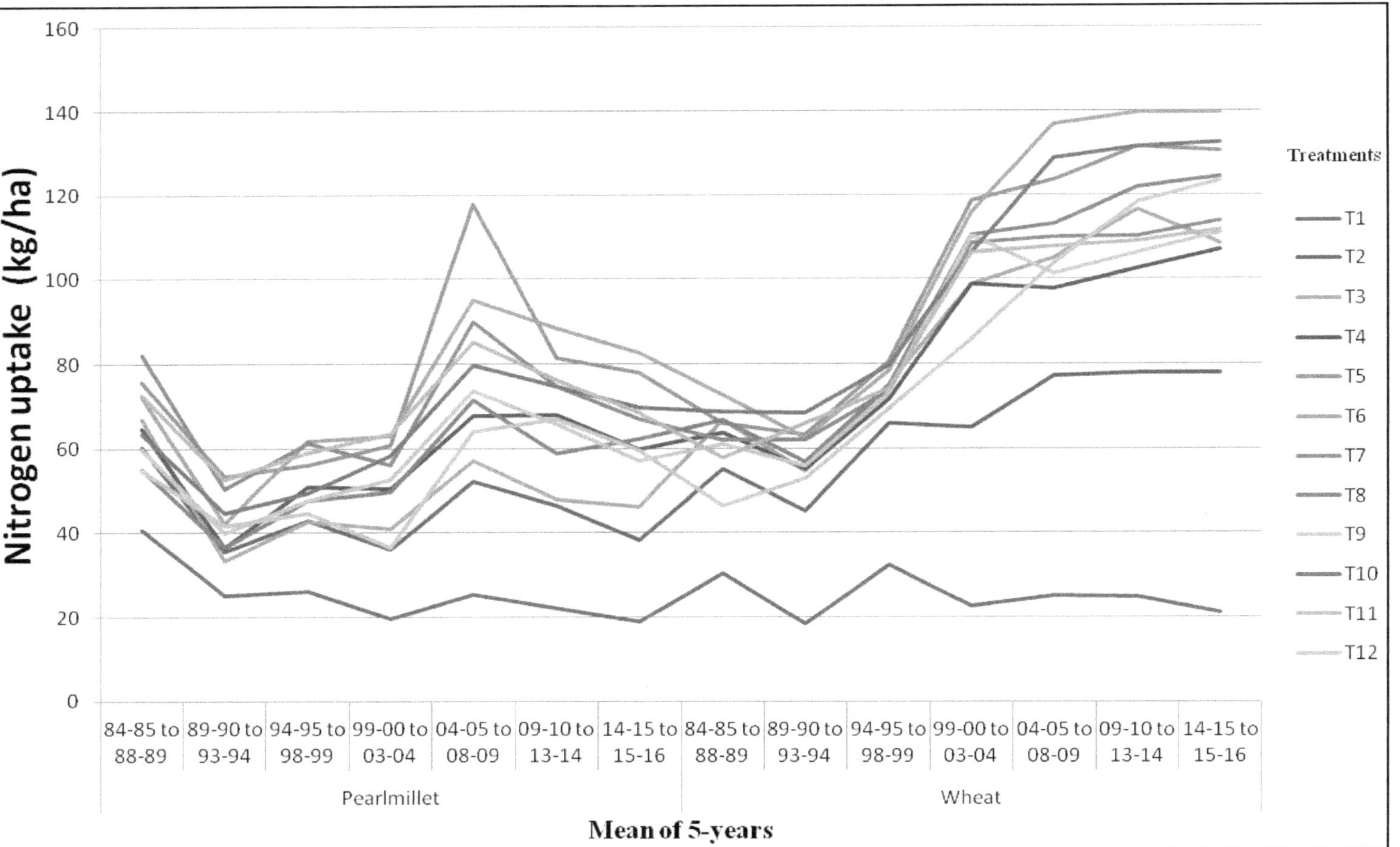

Figure 2: Nitrogen Uptake by Pearl Millet and Wheat in Pearl Millet-Wheat Cropping System (Means of 5-years).

Table 5: Effect of INM Treatments on Phosphorus Uptake by Pearl Millet and Wheat in Pearl Millet-Wheat Cropping System (Means of 5-years)

Treatment	*Phosphorus Uptake by Pearl Millet (kg/ha)*							*Phosphorus Uptake by Wheat (kg/ha)*						
	84-85 to 88-89	*89-90 to 93-94*	*94-95 to 98-99*	*99-00 to 03-04*	*04-05 to 08-09*	*09-10 to 13-14*	*14-15 to 15-16*	*84-85 to 88-89*	*89-90 to 93-94*	*94-95 to 98-99*	*99-00 to 03-04*	*04-05 to 08-09*	*09-10 to 13-14*	*14-15 to 15-16*
T_1	16.8	7.1	5.6	7.0	5.2	4.9	4.3	7.6	5.1	6.8	4.8	5.1	4.2	3.7
T_2	21.5	9.4	8.7	13.5	12.6	11.2	10.1	14.9	10.9	14.0	13.6	15.6	14.0	14.9
T_3	20.7	10.6	9.1	15.8	13.6	12.4	11.4	15.4	12.0	15.8	18.5	21.0	20.5	19.6
T_4	21.7	10.3	10.9	16.8	17.0	16.3	14.9	14.8	12.8	15.7	17.4	18.9	18.6	19.1
T_5	30.6	13.3	12.2	20.2	22.5	21.4	20.0	15.9	14.2	18.0	20.9	24.3	23.9	23.6
T_6	29.0	12.9	14.0	20.4	22.8	23.8	22.2	17.1	15.6	18.1	21.0	27.0	26.2	23.4
T_7	30.8	13.3	14.3	20.3	20.7	19.1	17.2	14.8	14.6	16.5	19.9	25.0	20.3	20.5
T_8	22.4	11.1	10.9	18.3	17.1	16.2	16.3	16.0	14.4	17.1	18.4	24.4	22.8	21.9
T_9	20.3	12.7	13.2	20.9	17.0	18.3	16.5	14.2	12.7	16.0	19.1	21.6	19.4	19.4
T_{10}	23.0	12.8	10.7	20.5	19.6	20.5	19.2	16.7	14.0	17.7	20.9	24.9	24.2	24.1
T_{11}	26.9	13.4	12.5	20.7	19.3	21.0	17.9	14.5	13.6	16.2	18.0	23.7	20.2	20.7
T_{12}	19.6	11.6	9.0	14.5	16.0	17.2	16.0	11.1	11.3	14.8	16.1	21.6	20.4	20.6

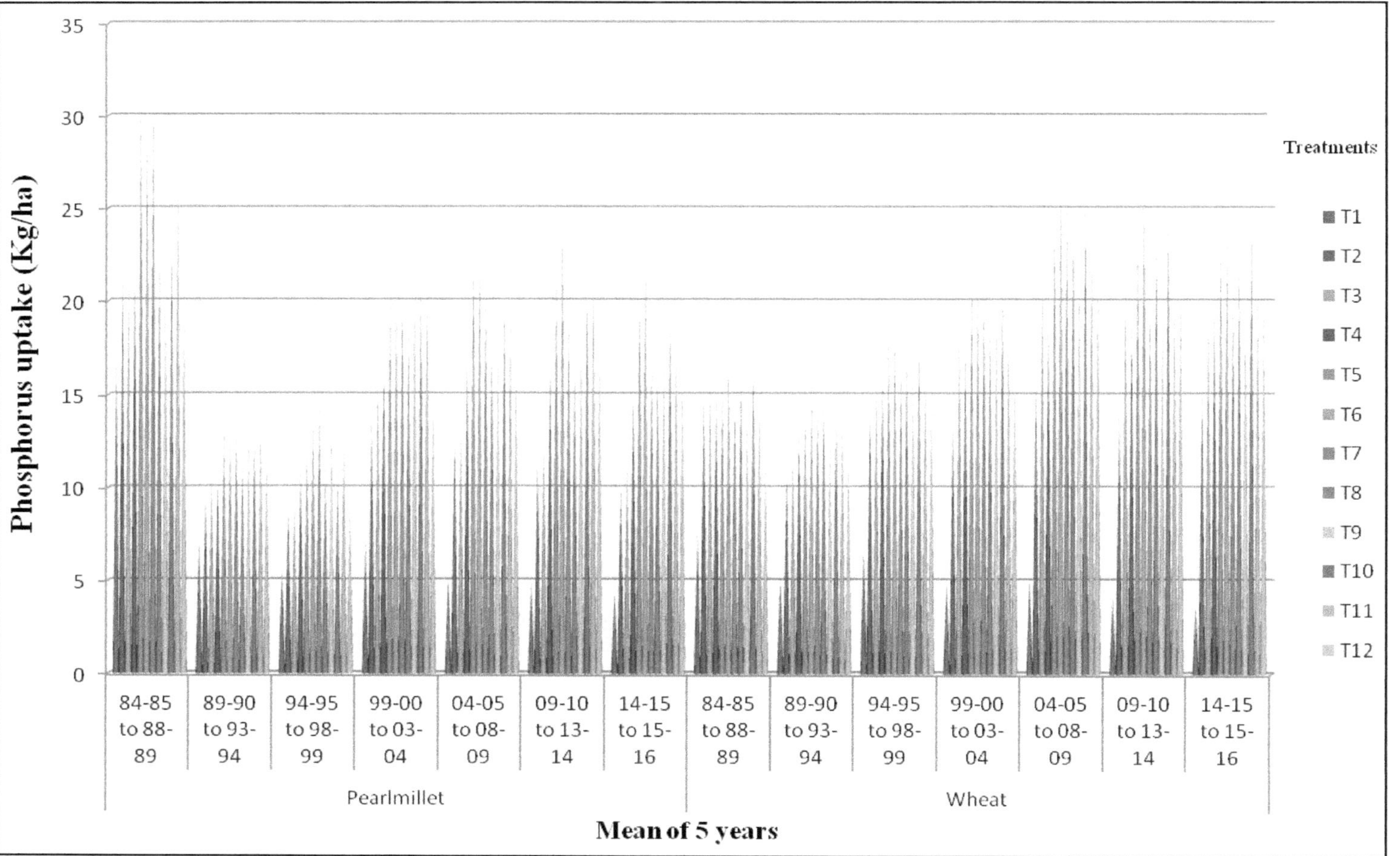

Figure 3: Phosphrous Uptake by Pearl Millet and Wheat in Pearl Millet-Wheat Cropping System (Means of 5-years).

where the increase observed was 45.5 per cent in the mean P uptake of 09-10 to 13-14 over the mean of 84-85 to 88-89. Similar pattern was observed during 2014-15 to 2015-16. Among organic sources, FYM had higher effect on crops productivity and thereby increased P uptake more as compared to other organic sources. Similar results of increase in P uptake with increased fertilizer dose have also been reported by Kumar *et al.*, 2005.

Potassium Uptake

Mean potassium uptake by pearl millet (Table 6) in control reduced to almost one-third on comparing the mean uptake of 09-10 to 13-14 with the mean of 84-85 to 88-89. In the treatments T_2, T_3, T_4 and T_5 where only inorganic fertilizers were used both during *Kharif* and *Rabi* season, respective, per cent decline in mean K uptake of 55.1, 52.0, 36.6 and 41.9 was observed. Integrated nutrient management (INM) treatments (T_6 to T_{11}) showed per cent variation ranged between 21.2 in treatment T_6 and 38.3 in treatment T_7. Per cent variation with respect to mean K uptake by pearl millet was almost at par both in treatments T_8 and T_{11} (31.0).

Mean potassium uptake by wheat (Table 6) in control showed reduction from 78.6 to 27.1 kg ha^{-1} as we move from the mean of 84-85 to 88-89 to the mean of 09-10 to 13-14. Among the treatments where only inorganic fertilizers were used *i.e.* treatments from T_2 to T_5, respective, per cent reduction of 41.0, 28.6, 40.6 and 13.4 was observed in mean potassium (K) uptake. Similar pattern was observed during 2014-15 to 2015-16. Among the INM treatments from T_6 to T_{11}, where different combinations of inorganic fertilizers and organic manures were used, respective, per cent reduction of 13.0 in treatment T_6 and 37.1 in treatment T_{11} was observed with respect to mean K uptake over the years. Similar results have also been reported by Laxminaryana and Patiram, 2006 that the incorporation of organics with chemical fertilizers significantly enhanced the nutrient uptake even higher than the optimum doses of NPK application. The increase in potassium uptake in pearl millet and wheat with the application of fertilizers/manures may be due to the reason that they increased cell division and elongation and simultaneously increased enzymatic activity and thereby resulted in more uptake of potassium.

Soil Fertility

Per cent Organic Carbon

The application of inorganic fertilizers alone (T_5) or in combination with FYM (T_6) in a long term fertilizer experiment resulted in maximum mean (09-10 to 13-14) per cent organic carbon (0.47) after the harvest of pearl millet. Per cent increase in organic carbon in the mean of 09-10 to 13-14 with respect to the mean of 84-85 to 88-89 in treatments T_5 and T_6 was 30.6 and 11.9, respectively. Whereas per cent decrease in organic carbon in control was 10.3. Increase or decrease (Table 7) in soil organic carbon was marginal on comparing the means of 84-85 to 88-89 with the means of 09-10 to 13-14. Similar pattern was observed during 2014-15 to 2015-16. However, recommended dose of NP fertilizers were tended to have more organic C in soil compared to half recommended dose of NP fertilizers. The higher increase in T_5 and T_6 treatments may be due to better crop growth with concomitant higher

Table 6: Effect of INM Treatments on Potassium Uptake by Pearl Millet and Wheat in Pearl Millet-Wheat Cropping System (Means of 5-years)

Treatment	*Potassium Uptake by Pearl Millet (kg/ha)*							*Potassium Uptake by Wheat (kg/ha)*						
	84-85 to 88-89	*89-90 to 93-94*	*94-95 to 98-99*	*99-00 to 03-04*	*04-05 to 08-09*	*09-10 to 13-14*	*14-15 to 15-16*	*84-85 to 88-89*	*89-90 to 93-94*	*94-95 to 98-99*	*99-00 to 03-04*	*04-05 to 08-09*	*09-10 to 13-14*	*14-15 to 15-16*
T_1	153.8	88.6	64.8	72.3	75.7	53.7	46.9	78.6	49.1	49.1	31.3	32.3	32.3	24.4
T_2	265.2	124.6	110.4	122.4	152.4	116.6	101.5	157.8	100.7	102.8	102.2	115.9	102.7	81.7
T_3	266.0	122.0	103.7	133.8	165.4	125.4	111.4	180.9	133.0	112.4	141.0	159.5	157.4	114.2
T_4	254.9	142.5	118.4	168.3	201.5	157.5	138.3	176.8	122.4	113.0	142.2	141.9	131.3	108.4
T_5	330.2	172.8	127.7	180.3	255.2	187.8	175.7	167.5	144.5	131.1	157.6	167.9	175.6	140.4
T_6	261.3	151.4	144.5	186.7	269.7	199.3	192.8	180.8	139.0	135.1	148.7	185.8	191.3	152.2
T_7	301.4	167.0	147.0	177.4	248.2	180.5	166.0	163.9	138.6	121.5	140.3	168.4	151.3	116.4
T_8	219.8	125.7	116.8	162.7	205.7	151.1	160.8	195.0	139.8	127.9	154.6	162.3	161.3	122.9
T_9	236.5	148.2	142.8	171.7	215.4	165.2	150.1	159.4	115.7	118.2	144.6	155.1	140.8	112.3
T_{10}	260.2	160.9	123.4	178.0	230.5	184.7	174.5	178.0	148.9	128.6	146.4	174.1	179.9	132.1
T_{11}	275.7	153.4	134.6	186.8	237.0	187.2	172.3	169.3	134.0	116.4	142.7	159.9	136.0	58.2
T_{12}	188.9	134.6	106.5	119.9	186.8	166.5	151.3	129.6	97.3	106.8	118.8	148.9	156.0	125.3

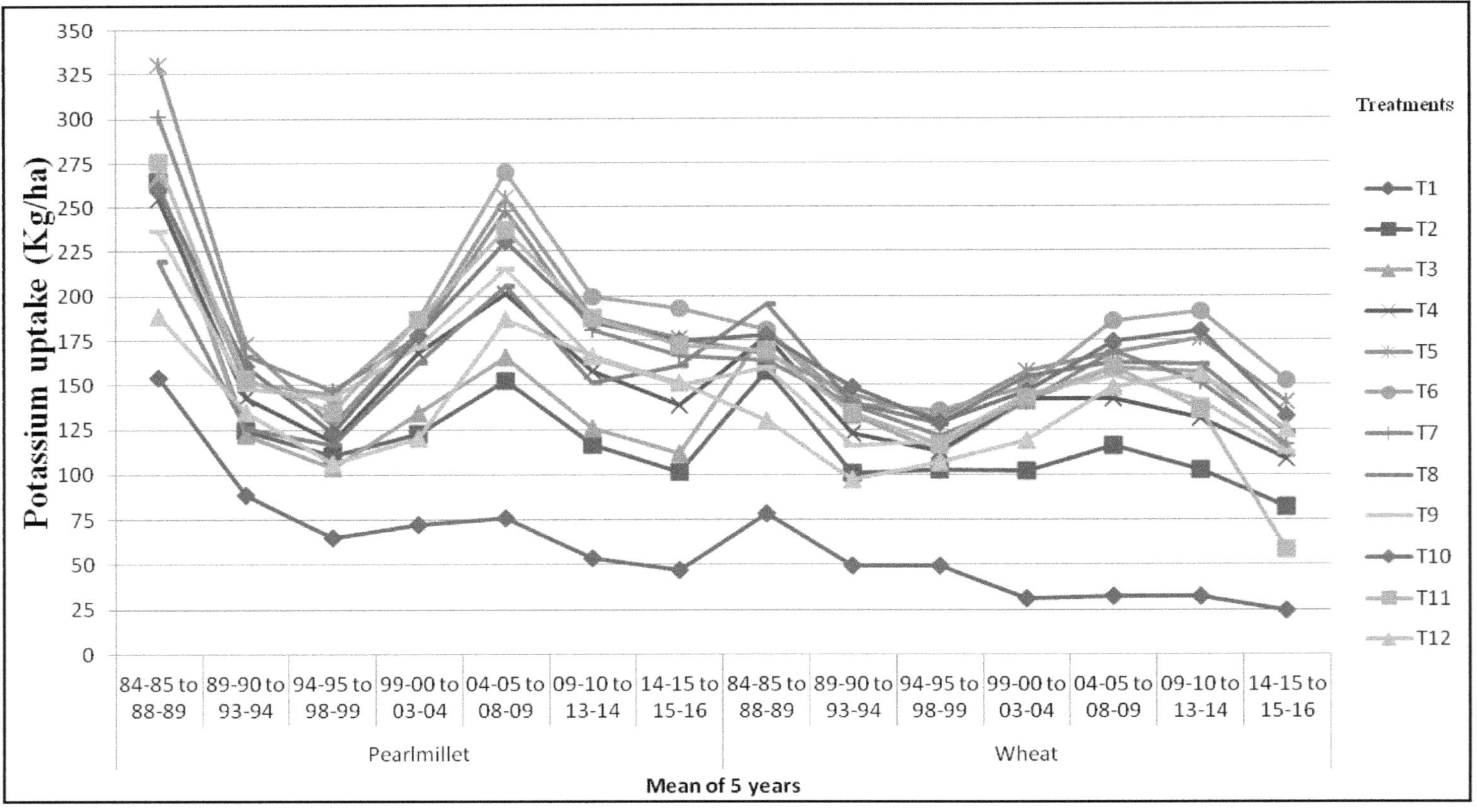

Figure 4: Potassium Uptake by Pearl Millet and Wheat in Pearl Millet-Wheat Cropping System (Means of 5-years).

root biomass generation and higher return of left over surface plant residues. It may be postulated that more production of roots and their subsequent decomposition increased organic carbon content in soil (Antil and Mandeep (2007); Antil and Narwal (2007).

After the harvest of wheat crop (Table 7) almost similar type of results were obtained as after the harvest of pearl millet among different treatments. Per cent increase in organic carbon in the mean of 09-10 to 13-14 with respect o the mean of 84-85 to 88-89 in treatments T_5 and T_6 was 6.8 and 11.1, respectively. Similar pattern was observed during 2014-15 to 2015-16. The per cent decrease in organic carbon in control was 14.6. Reasons for the marginal changes in per cent organic carbon among rest of the treatments were same as after the harvest of pearl millet crop.

Application of organic amendments as supplements with NPK not only added organic carbon in the soil but also increased plant C inputs in the soil through root residue, stubble, rhizodeposition etc. because of an increase in yield. Kukal *et al.* (2009) also observed a higher C sequestration in a 33-year old rice-wheat system due to application of FYM and the cropping system has greater capacity to sequest C because of high C input through enhanced productivity. A 20-40 per cent increase in organic C with continuous use of FYM was also reported by Ruhal and Shukla (1979).

Available Nitrogen

The available (mineralizable) nitrogen (N) content in soil (Table 8) after the harvest of pearl millet crop decreased from 148.5 kg ha^{-1} (mean of 84-85 to 88-89) to 129.1 kg ha^{-1} (mean of 09-10 to 13-14). A marked increase in available N in the mean of 09-10 to 13-14 with respect to the mean of 84-85 to 88-89 in treatments T_6 and T_9 where organic manures (FYM and wheat straw) were applied in conjunction with NP fertilizers. The respective per cent increase in available N in treatments T_6 and T_9 was 32.8 and 18.8. Whereas, per cent increase in treatment T_5 where 100 per cent recommended dose of chemical fertilizers were applied both during *Kharif* and *Rabi* seasons was 6.8 only. Increase in available N content with the addition of organic manures might be due to the release of N through the decomposition of organic manures (Antil and Mandeep (2007); Antil and Narwal (2007). Moreover, it could also be attributed to the greater multiplication of soil microbes, which could convert organically bound N into inorganic form. Among the different organic manures, FYM registered more increase in available N content of soil as compared to wheat straw and green manure. Some microbes may be N limited if C is in excess supply, leading to variation in the abundance and proportions of fungi and bacteria and thus differences in soil N cycling (Boyle *et al.*, 2008). Organic manures containing N that are released into soils during litter degradation will remain in soil solution only transitionally becoming bound and precipitated by high molecular weight polyphenolic compounds which are generated from applied organic manures due to larger N availability in soil (Majuakim and Kitayana (2013).

After the harvest of wheat crop (Table 8), marked increase in available N in soil was noticed in the means of 09-10 to 13-14 over the means of 84-85 to 88-89 among all treatments except in control where 12.6 per cent decrease in available N was noticed. Comparison of the means of 09-10 to 13-14 with the means of 84-85 to 88-89

Table 7: Effect of INM Treatments on Per cent Organic Carbon after Pearl Millet and Wheat in Pearl Millet-Wheat Cropping System (Means of 5-years)

Treatment	*Per cent Organic Carbon after Pearl Millet*							*Per cent Organic Carbon after Wheat*						
	84-85 to 88-89	*89-90 to 93-94*	*94-95 to 98-99*	*99-00 to 03-04*	*04-05 to 08-09*	*09-10 to 13-14*	*14-15 to 15-16*	*84-85 to 88-89*	*89-90 to 93-94*	*94-95 to 98-99*	*99-00 to 03-04*	*04-05 to 08-09*	*09-10 to 13-14*	*14-15 to 15-16*
T_1	0.39	0.40	0.38	0.42	0.40	0.35	0.35	0.41	0.51	0.48	0.56	0.41	0.35	0.36
T_2	0.38	0.35	0.39	0.41	0.45	0.38	0.40	0.44	0.51	0.46	0.56	0.45	0.40	0.41
T_3	0.37	0.38	0.44	0.47	0.46	0.41	0.41	0.45	0.49	0.51	0.59	0.54	0.44	0.44
T_4	0.40	0.36	0.42	0.46	0.48	0.42	0.41	0.42	0.49	0.45	0.56	0.50	0.44	0.44
T_5	0.36	0.33	0.45	0.47	0.51	0.46	0.48	0.44	0.47	0.42	0.54	0.52	0.47	0.50
T_6	0.42	0.42	0.47	0.53	0.55	0.47	0.50	0.45	0.56	0.48	0.58	0.60	0.50	0.51
T_7	0.45	0.40	0.38	0.50	0.54	0.44	0.43	0.48	0.53	0.50	0.58	0.57	0.48	0.46
T_8	0.43	0.52	0.42	0.58	0.52	0.42	0.42	0.44	0.68	0.53	0.62	0.55	0.46	0.44
T_9	0.43	0.45	0.45	0.53	0.50	0.41	0.41	0.46	0.59	0.47	0.62	0.56	0.44	0.43
T_{10}	0.41	0.43	0.40	0.46	0.51	0.43	0.45	0.48	0.51	0.53	0.56	0.53	0.46	0.47
T_{11}	0.41	0.40	0.44	0.44	0.51	0.42	0.44	0.43	0.49	0.51	0.58	0.57	0.44	0.46
T_{12}	0.37	0.41	0.46	0.47	0.46	0.41	0.42	0.45	0.48	0.50	0.57	0.51	0.41	0.43

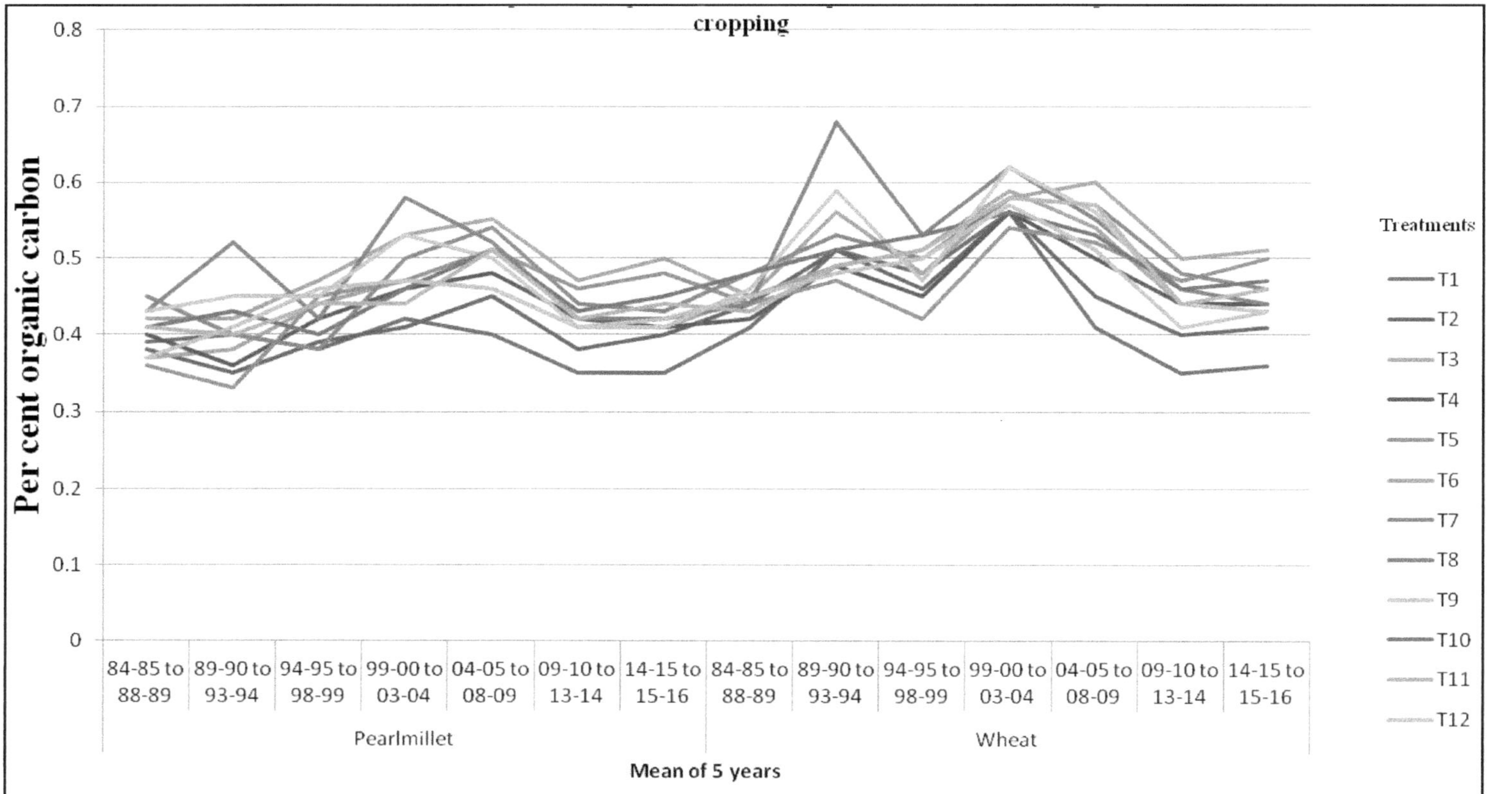

Figure 5: Organic Carbon (Per cent) after Pearl Millet and Wheat in Pearl Millet-Wheat Cropping System (Means of 5-years).

Table 8: Effect of INM Treatments on available N after Pearl Millet and Wheat in Pearl Millet-Wheat Cropping (Means of 5-years)

Treatment	Available Nitrogen (N) after Pearl Millet (kg ha^{-1})							Available Nitrogen (N) after Wheat (kg ha^{-1})						
	84-85 to 88-89	89-90 to 93-94	94-95 to 98-99	99-00 to 03-04	04-05 to 08-09	09-10 to 13-14	14-15 to 15-16	84-85 to 88-89	89-90 to 93-94	94-95 to 98-99	99-00 to 03-04	04-05 to 08-09	09-10 to 13-14	14-15 to 15-16
T_1	148.5	102.6	97.4	115.9	122.9	128.8	132.2	165.2	130.1	112.1	124.7	159.3	142.8	142.6
T_2	158.1	99.6	92.9	114.6	148.8	148.4	147.9	164.3	129.5	115.8	147.3	172.0	162.8	163.65
T_3	161.8	109.6	91.1	132.2	157.5	156.8	165.4	140.8	120.4	122.4	145.6	193.6	173.4	178.55
T_4	148	109.2	98.7	120.1	162.1	169.4	173.3	157.9	118.6	115.1	156.3	184.9	189.3	189
T_5	178.6	113.9	94.7	129.4	175.4	190.8	196.9	158.5	120.4	122.4	171.9	223.7	213.1	209.15
T_6	148.7	107.5	100.5	136.5	182.0	197.1	202.2	167.0	130.1	119.7	180.3	233.5	220.2	212.7
T_7	155	105.6	86.2	132.9	175.6	184.5	181.2	147.2	132.2	124.6	148.0	209.3	207.2	198.6
T_8	160.1	106.0	98.9	120.2	158.2	175.0	169.8	163.7	119.8	119.9	152.3	215.3	197.1	192.6
T_9	151.3	105.3	103.1	135.7	165.2	177.8	169.8	164.4	121.4	115.9	172.9	205.5	196.3	192.5
T_{10}	171.9	106.4	90.9	118.7	165.6	180.6	190.8	152.3	123.0	110.6	172.3	208.3	204.8	207.5
T_{11}	167.7	134.1	89.5	130.0	167.3	177.8	176.8	166.8	126.5	115.3	165.9	204.4	204.1	196.9
T_{12}	167.3	124.0	88.7	119.8	147.7	171.5	165.4	173.6	126.9	108.3	151.7	189.0	197.1	191.65

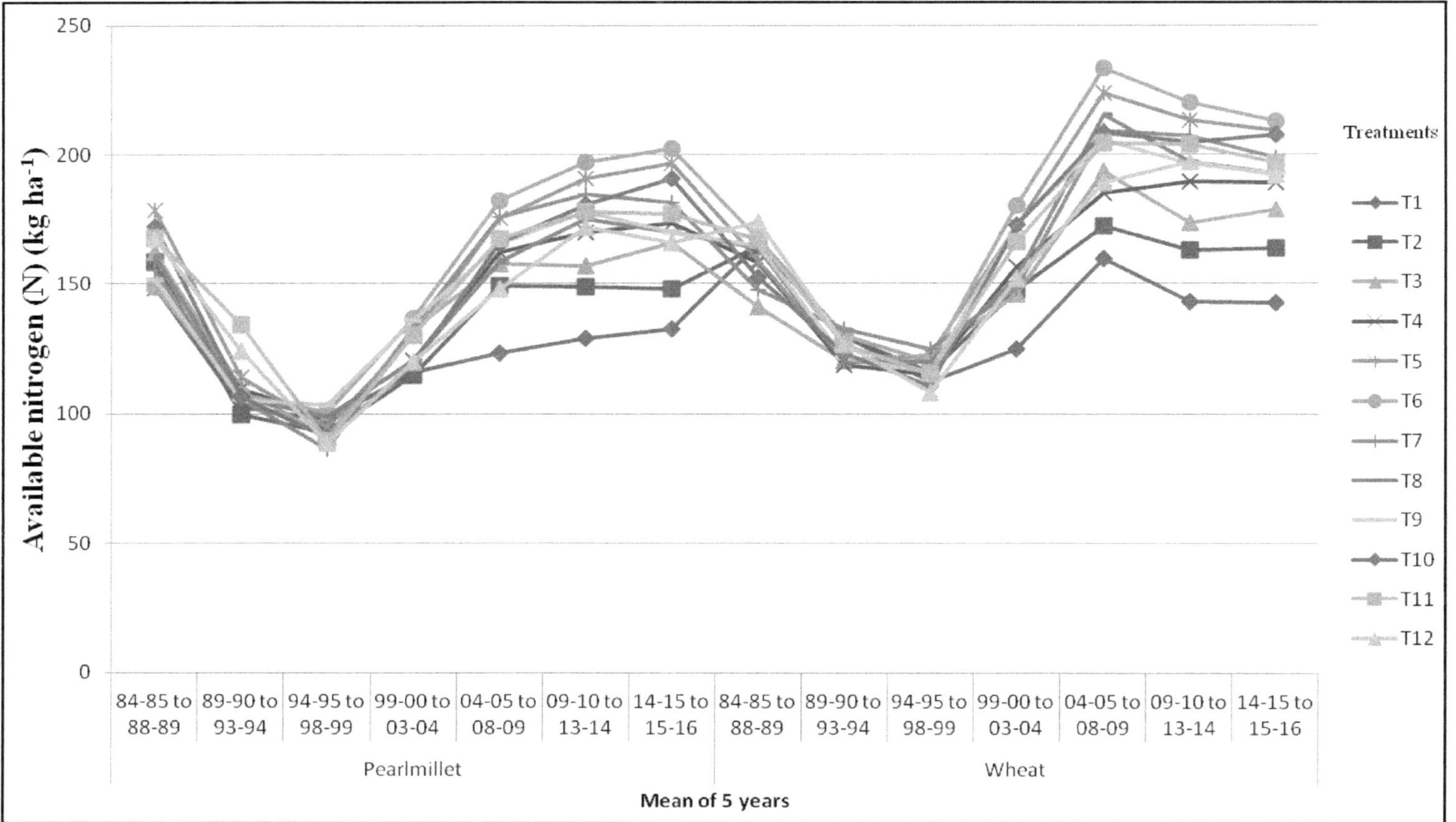

Figure 6: Available N after Pearl Millet and Wheat in Pearl Millet-Wheat Cropping System (Means of 5-years).

showed no variation in available N in treatment T_2 where 50 per cent recommended dose of fertilizers were added during both *Kharif* and *Rabi* seasons. Per cent increase in available N varied between 19.5 in treatment T_9 and 40.3 in treatment T_7 among different treatments. Per cent increase (35.6) each was at par in treatments T_5 and T_{10}. Similar pattern was observed during 2014-15 to 2015-16.

Available Phosphorus

After the harvest of pearl millet crop, mean available phosphorus (P) content (kg ha^{-1}) (Table 9) of the soil in control decreased (28 per cent) from 18.8 kg ha^{-1} (mean of 84-85 to 88-89) to 13.4 kg ha^{-1} (mean of 09-10 to 13-14). Per cent decrease in mean available P status on comparing the means of 84-85 to 88-89 with the means of 09-10 to 13-14 ranged between 3.1 in farmers' practice and 35.0 in treatment T_9. Treatment T_9 was followed by treatment T_8 where 30.2 per cent decrease in available P was observed. Among treatments, where only inorganic fertilizers were used, per cent decrease in available P ranged between 15.3 and in treatment T_2 and 28.1 in T_3. Among treatments T_6 to $T_{11,}$ where different combinations of organic and inorganic fertilizers were used, per cent decrease in available P ranged between 11.1 in T_{10} and 35.0 in T_9. However, a glance at the data within the treatments of means of 5- years each shows that mean available P in soil was better in treatments containing combinations of various organic manures and inorganic fertilizers used. On comparing means of 84-85 to 88-89 with 09-10 to 13-14, per cent reduction in mean available P was around 16.0 both in treatments T_5 and T_6 and 11.1 in treatment T_{10}. In totality treatments involving combination of various organic manures and inorganic fertilizers besides treatment where recommended dose of inorganic fertilizers was added both during both seasons, mean available P status sustained better over the years than in rest of the treatments.

Similarly after the harvest of wheat crop (Table 9), per cent decrease in mean available P in soil ranged between 10.2 in treatment T_5 where 100 per cent recommended dose of fertilizers was added during both *Kharif* and *Rabi* seasons. Among combinations of various organic manures and inorganic fertilizers, mean available P ranged between 10.4 and 36.4 per cent in treatments T_{11} and T_8, respectively. Across means of 5-years data within the treatments mean available P status was always better in treatment T_6 than rest of the treatments. Mean available P in treatment T_6 performed better than rest of the combination of organic manures and inorganic fertilizers across most of the means of 5-years data. Per cent reduction in means of 84-85 to 88-89 with the means of 09-10 to 13-14, on comparison showed minimum reduction in T_6 treatment *i.e.* from 23.3 to 19.8 kg ha^{-1} *i.e.* 15 and in treatment T_5 where 100 per cent recommended dose of fertilizers was added both during *Kharif* and *Rabi* seasons was 10.2 and in T_{11} it was 10.4. So these treatments *i.e.* T_5, T_6 and T_{10} sustained mean available P status of soil both in the beginning and at the end of the experiment.

Available Potassium

Comparing means of available potash 84-85 to 88-89 with that of 09-10 to 13-14, (K), per cent decrease was observed maximum (52.6) in control after the harvest of pearl millet (Table 10). Among different inorganic treatments it varied between

Table 9: Effect of INM Treatments on available P after Pearl Millet and Wheat in Pearl Millet-Wheat Cropping System (Means of 5-years)

Treatment	Available Phosphorus (P) after Pearl Millet (kg ha^{-1})							Available Phosphorus (P) after Wheat (kg ha^{-1})						
	84-85 to 88-89	89-90 to 93-94	94-95 to 98-99	99-00 to 03-04	04-05 to 08-09	09-10 to 13-14	14-15 to 15-16	84-85 to 88-89	89-90 to 93-94	94-95 to 98-99	99-00 to 03-04	04-05 to 08-09	09-10 to 13-14	14-15 to 15-16
T_1	18.8	14.5	15.3	13.1	10.4	13.5	14.3	17.4	17.3	12.2	24.4	8.4	12.5	15.0
T_2	17.0	14.2	16.5	13.8	14.5	14.5	15.0	26.5	24.5	15.1	28.3	11.1	13.8	16.0
T_3	19.2	18.2	17.7	13.6	17.8	14.2	16.8	19.1	21.9	14.5	29.8	14.3	15.1	16.5
T_4	18.7	33.3	21.9	14.2	16.6	16.0	18.7	24.2	25.0	14.1	27.3	13.8	16.4	18.2
T_5	22.2	20.4	15.4	17.1	19.8	19.0	20.3	20.6	21.3	15.8	35.6	17.2	18.5	19.4
T_6	23.3	33.4	19.3	18.5	18.8	19.8	20.9	23.3	28.0	20.6	40.0	19.1	19.8	20.9
T_7	22.9	26.5	19.5	19.6	17.2	17.9	20.7	21.1	27.8	17.6	37.2	17.3	17.7	18.8
T_8	22.2	24.5	18.8	17.0	19.6	15.6	17.8	25.0	20.9	18.8	32.4	15.0	16.1	18.7
T_9	22.3	20.6	19.6	17.6	18.0	15.0	18.3	21.0	19.7	16.2	30.2	14.6	17.0	18.0
T_{10}	19.8	17.0	20.0	14.9	19.9	18.2	19.7	23.1	19.3	16.9	31.9	13.6	17.0	19.4
T_{11}	22.7	18.6	19.7	13.8	20.1	17.6	18.8	20.1	17.3	17.5	34.8	15.3	18.0	18.3
T_{12}	15.9	13.6	17.4	15.5	17.4	15.7	16.5	16.6	12.6	16.3	28.6	13.1	16.1	18.5

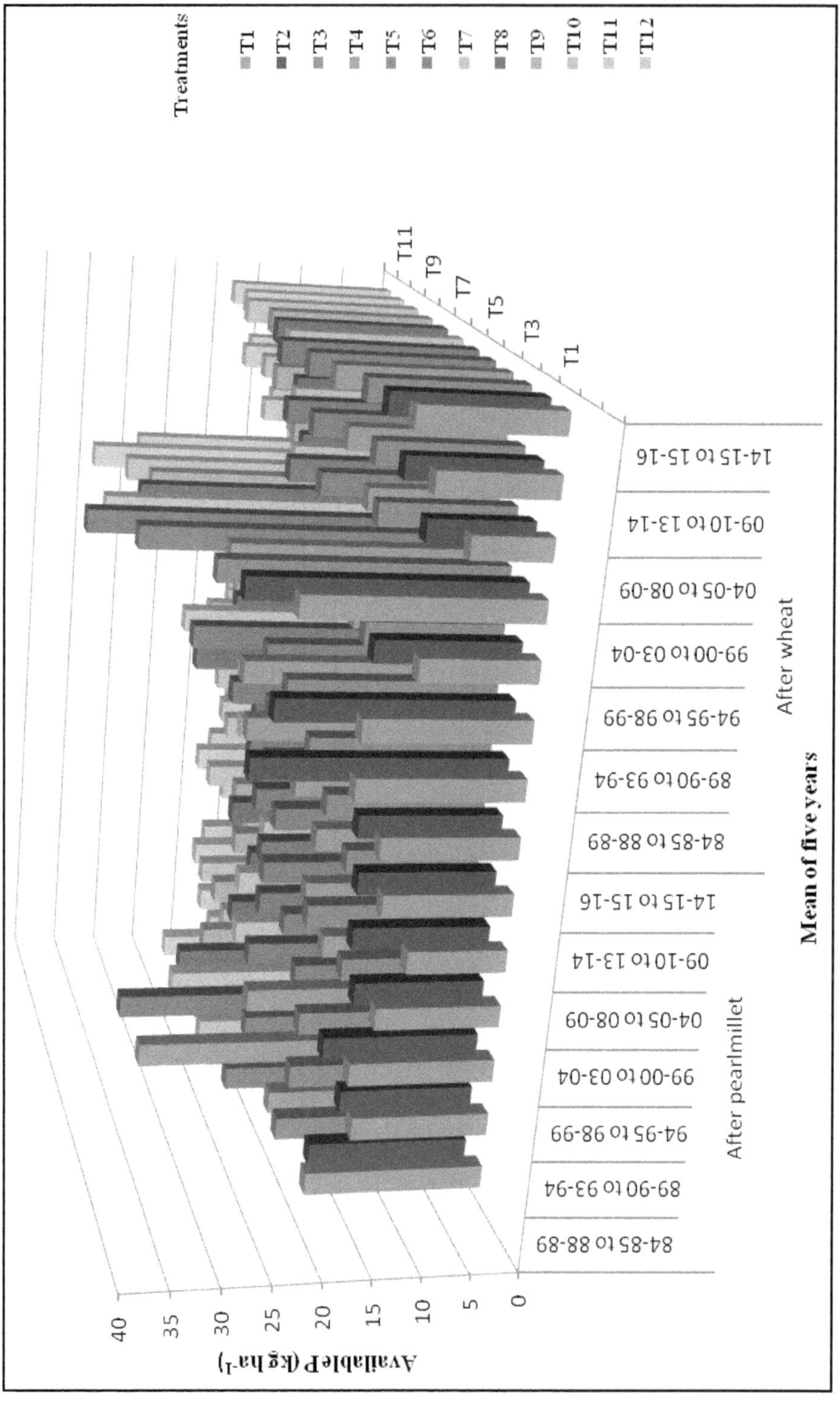

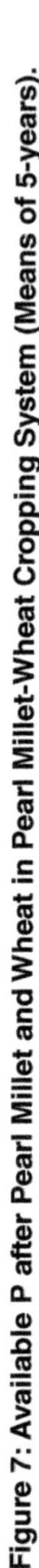

Figure 7: Available P after Pearl Millet and Wheat in Pearl Millet-Wheat Cropping System (Means of 5-years).

Table 10: Effect of INM Treatments on available K after Pearl Millet and Wheat in Pearl Millet-Wheat Cropping System (Means of 5-years)

Treatment	Available potassium (K) after Pearlmillet (kg ha^{-1})							Available potassium (K) after wheat (kg ha^{-1})						
	84-85 to 88-89	89-90 to 93-94	94-95 to 98-99	99-00 to 03-04	04-05 to 08-09	09-10 to 13-14	14-15 to 15-16	84-85 to 88-89	89-90 to 93-94	94-95 to 98-99	99-00 to 03-04	04-05 to 08-09	09-10 to 13-14	14-15 to 15-16
T_1	466.9	455.9	278.0	298.8	222.0	223.2	232.0	457.5	416.4	316.1	316.9	208.0	207.1	214.4
T_2	443.1	391.4	257.0	267.0	256.5	242.4	240.5	456.5	353.7	292.7	284.0	240.5	225.1	225.7
T_3	429.3	398.1	262.7	288.5	279.8	264.3	250.0	448.3	410.2	289.8	298.1	258.0	252.5	258.1
T_4	402.1	402.2	228.3	257.9	290.8	285.5	264.5	441.1	370.3	308.3	300.5	274.0	261.7	259.6
T_5	402.1	368.9	224.0	259.1	287.5	302.2	296.3	415.7	349.2	306.2	276.9	308.3	304.6	299.0
T_6	481.2	468.1	250.5	287.4	293.8	307.4	298.8	472.9	394.4	315.6	302.2	326.5	312.7	302.8
T_7	401.5	448.2	243.5	282.3	308.3	297.2	291.3	482.9	371.6	312.7	313.2	315.8	305.9	294.3
T_8	450.1	602.0	256.3	375.5	319.5	287.6	272.2	500.3	487.2	373.8	367.8	311.3	297.7	286.6
T_9	497.9	499.2	278.5	312.1	339.5	283.8	268.3	455	432.4	310.7	327.1	314.5	297.6	282.7
T_{10}	426.6	418.9	244.6	275.1	300.5	294.5	290.6	494.96	364.8	305.4	299.7	303.8	303.3	293.7
T_{11}	453.2	425.7	216.7	254.1	298.0	284.9	268.1	457.04	340.9	282.7	284.7	326.5	294.7	290.8
T_{12}	416.4	401.2	209.0	250.1	271.5	276.7	244.5	440.1	333.6	278.0	292.0	278.0	286.8	281.0

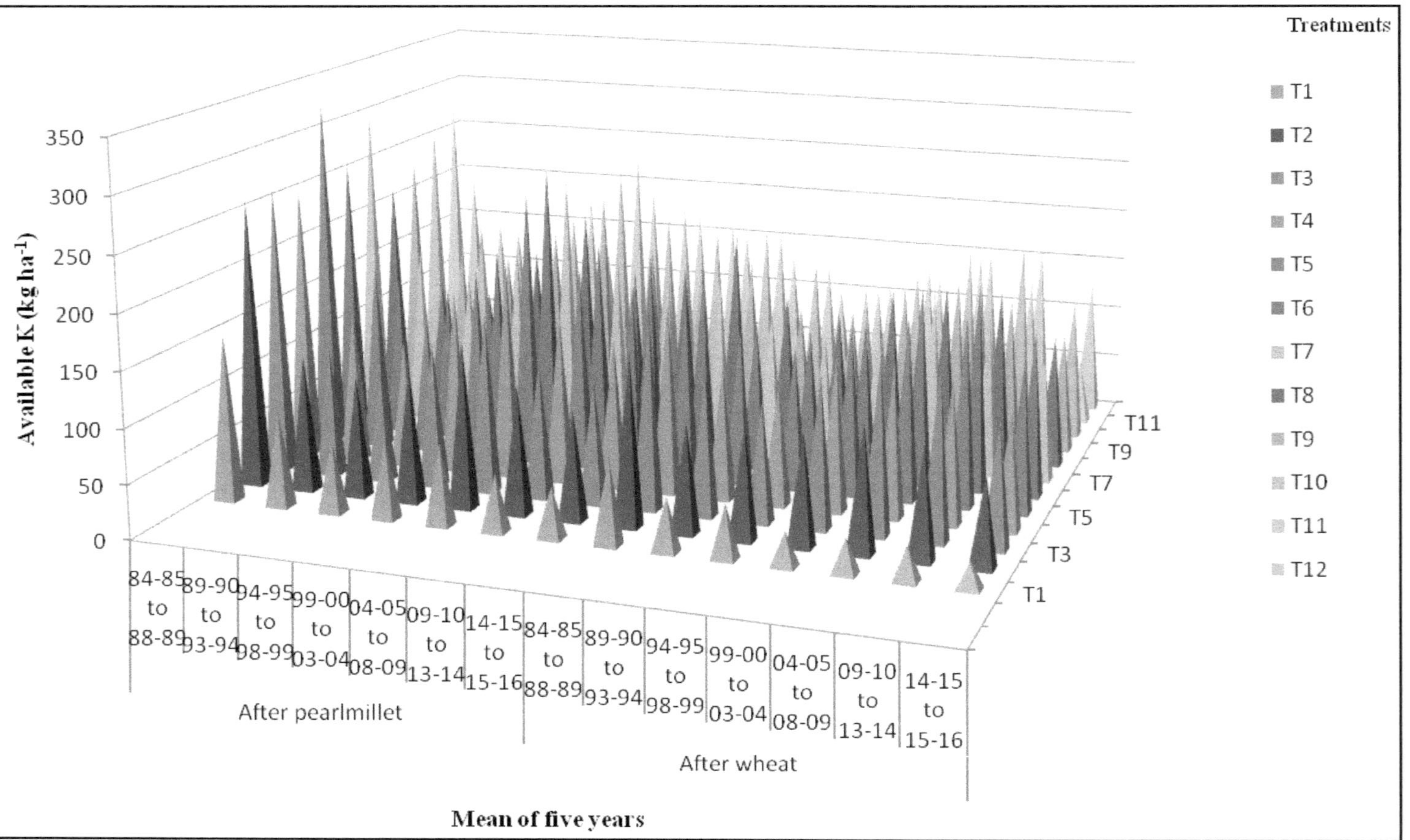

Figure 8: Available K after Pearl Millet and Wheat in Pearl Millet-Wheat Cropping System (Means of 5-years).

24.7 in treatment T_5 where 100 per cent recommended dose of fertilizers was added both during *Kharif* and *Rabi* seasons and 45.2 in treatment T_2 where 50 per cent recommended dose of fertilizers was added during both *Kharif* and *Rabi* seasons. So we can say that treatment T_5 sustained better mean available K status than rest of the inorganic treatments. Among combinations of different organic manures and inorganic fertilizers, treatment T_7 showed maximum sustainability *i.e.* 25.8 per cent reduction with respect to mean available K after the harvest of pearl millet crop followed by treatment T_{10} with 31.0 per cent sustainability. Minimum sustainability with respect to available K *i.e.* maximum reduction of 40.1 per cent was observed in treatment T_8. Similarly among inorganic treatments after the harvest of wheat crop (Table 10); per cent reduction in the means of 84-85 to 88-89 with respect to the mean of 09-10 to 13-14 was observed minimum (26.2) in treatment T_5 and maximum (50.4) in T_2. Mean available K reduction was observed minimum *i.e.* about 33.1 per cent in treatment T_6 among different combinations of various organic manures and inorganic fertilizers on comparing means of 84-85 to 88-89 with respect to the means of 09-10 to 13-14 and maximum reduction in the treatment T_8 with 40.1 per cent. Similar pattern was observed during 2014-15 to 2015-16. Verma *et al.* (1987) also reported decrease of 48.8 per cent in control plot and 18.9 per cent in plots receiving 40 kg N and 60 kg P_2O_5 ha^{-1}.

Conclusion

During initial years of experimentation, FYM application for 25 per cent N requirement gave higher grain yield of pearl millet 3.87, 20.82, 22.10, 15.28 and 2.73 per cent, respectively over 50 per cent N from FYM, 50 per cent N through wheat straw, 25 per cent N through wheat straw, 50 per cent N through green manure and 25 per cent N through green manure. 100 per cent recommended NPK produced 6.77 per cent higher yield over where 25 per cent N was replaced with FYM and 75 per cent NPK was applied through chemical fertilizer in pearl millet and 75 per cent NPK was applied in wheat. During this period 53.28 per cent low yield was produced over 100 per cent recommended NPK to both the crops. After about 15 years of experimentation, the increase in yield was 401, 554 and 752 kg/ha with fertilization of 50 per cent N through FYM in pearl millet where grain yield was recorded 2584 kg/ha. At this stage of experimentation, the yields were almost same (2584 and 2586 kg/ha) with the substitution of N through FYM to the tune of 50 per cent and 25 per cent. Earlier, 25 per cent N substitution was giving high yield over 50 per cent N through FYM. Over the years, slightly higher wheat grain yield was recorded with the application of 50 per cent N through FYM + 50 per cent recommended NPK through chemical fertilizers in pearl millet, followed by 100 per cent recommended NPK in wheat. Among FYM, wheat straw and green manure application to substitute 50 per cent N in pearl millet along with chemical fertilizers, FYM performed better followed by green manure. The WEY from 2002-03 to 2015-16 was higher where 50 per cent N was substituted through FYM + 50 per cent recommended NPK through chemical fertilizers in pearl millet followed by 100 per cent recommended NPK though chemical fertilizers in wheat. Starting from 2002-03 to 2003-04 and towards the end of experimentation *i.e.* 2010-11 to 2015-16, the return over variable cost/net returns were higher every year with 50 per cent N application through FYM in pearl millet except during 2002-03 to 2003-04. With no application of fertilizer, the pearl millet-wheat cropping system was not remunerative.

Per cent increase in organic carbon in the mean of 09-10 to 12-13 with respect to the mean of 84-85 to 88-89 in treatments T_5 and T_6 was 30.6 and 11.9, respectively. Whereas per cent decrease in organic carbon in control was 10.3. After the harvest of pearl millet crop, mean available phosphorus (P) content (kg ha^{-1}) of the soil in control decreased (28 per cent) from 18.8 kg ha^{-1} (mean of 84-85 to 88-89) to 13.4 kg ha^{-1} (mean of 09-10 to 12-13). Per cent decrease in mean available P status on comparing the means of 84-85 to 88-89 with the means of 09-10 to 12-13 ranged between 3.1 in farmers' practice and 35.0 in treatment T_9. Similarly, after the harvest of wheat crop per cent decrease in mean available P in soil ranged between 10.2 in treatment T_5 where 100 per cent recommended dose of fertilizers was added during both *Kharif* and *Rabi* seasons. It can be concluded that treatment T_5 sustained better mean available K status than rest of the inorganic treatments both after the harvest of pearl millet and wheat.

Future Line of Work

Enhancing cropping systems' productivity, profitability, cost effectiveness and improving quality of food is a challenge to the scientists. Technological challenges are becoming more complex than before as demand for food is increasing, land holding size is decreasing and natural resource base is shrinking. For this a change in mind set towards intensive cropping systems along with practicing integrated nutrient management is needed to accomplish the mission of evergreen revolution.

Acknowledgement

The authors feel highly indebted to the Indian Institute of Farming Systems Research, ICAR, Modipuram for providing technical guidance and financial support to carry out this experiment for such a long time (since 1984-85) in the Department of Agronomy, CCS Haryana Agricultural University, Hisar. The authors also thankfully acknowledge the contribution of scientific staff worked in the project since 1984-85 to till date.

Chapter 7
Organic Farming

Development of Organic Farming Package for Mungbean-Wheat (*Desi*) Cropping System

The increasing awareness of the deleterious effects of indiscriminate use of chemical fertilizers and pesticides in agriculture has led to the adoption of organic farming as an alternative method for conventional farming world-wide. It is gaining gradual momentum across the world and the growing awareness of organic food which is emerging as an attractive source of rural income generation.

Haryana is an agriculturally advance state of India where in rice-wheat, cotton-wheat and pearl millet-wheat cropping systems are most prevalent. But now, these cropping systems are not proving farmers' friendly due to higher water requirement, declining yield, susceptibility to insects and pests and adverse effect on soil health (Chakrabarti and Singh, *2004*). Moreover, in the new millennium with the growing demand of agricultural commodities in national as well as international market due to ever increasing population and globalization. Indian farmers are forced to produce higher quantum of quality agricultural commodities at low cost from shrinking land and natural resources.

Mungbean - wheat cropping system requires less water and hence, can be grown in areas where water availability is limited in addition to beneficial effect of mungbean to supply N to the succeeding wheat crop which are favorable for organic farming (Gangaiah *et al., 2012*). Wheat (*desi*) also known as tall wheat requires less water and is very responsive to low doses of nutrients. The dwarf wheat grown by farmers with the application of chemicals in the form of fertilizers, herbicides, insecticides and pesticides is not preferred by people and therefore, the demand of organic wheat is rising fast in national and international market. Likewise, mungbean crop is relatively resistant/tolerant to dry land conditions in Haryana. The mungbean have deep penetrating tap root system which enables them to tap moisture from deep strata of the soil and can be easily grown with limited water in dry areas. By virtue of the quality, mungbean has found a niche in the areas characterized by moisture stress and therefore, mungbean is found very suitable crop for a sustainable cropping system. The short duration varieties of pulses like, mungbean help in increasing the cropping intensity and is a good catch crop. Suitable varieties for *Kharif* and summer season have opened new opportunities in increasing mungbean production. Mung bean constitutes the major part of human diet among pulses.

There is need to shift a portion of cultivable land with organic farming as an alternate practice technique. Organic farming gives emphasis on sustainability, environmental safety and low cost production techniques over chemical farming. Keeping the issues of 21st century agriculture in view following aspects also shows importance of organic farming: Reduce the toxic load; keep chemicals out of the air, water, soil and our bodies; reduce if not eliminate off farm pollution; protect future generations; build healthy soil; taste better and truer flavor; assist farmers of all sizes; promote biodiversity; celebrate the culture of agriculture etc.

Though after the advent of green revolution high yielding dwarf varieties of wheat are common with Haryana farmers. Yet the importance of growing wheat (*desi*) is still there. It is in view of its low cost of cultivation, low nutrient requirement, high nutritional value, high market price, low preference and adaptability to climate and soils in the area.

Organic Farming in Area/Zone/State

In general India and particularly Haryana and other northern states are traditionally organic in nature and in present agriculture along with FYM farmers have started growing pulses and legume forages which help in improving soil health and crop productivity. Green manuring has also been started in areas where rice-wheat cropping system is being followed for many years.

The areas adjoining Delhi, where farmers take high value crops which fetch better market price, organic farming is common. Organic farming is practiced in crops like vegetables, wheat (*desi*), basmati rice, pulses etc.

In south-west climatic zone of Haryana where soils are sandy loam having low organic matter, the mungbean – wheat (*desi*) cropping system is being practiced by the farmers. This cropping system is beneficial to the farmers as the nutrient requirement of wheat (*desi*) is low and mungbean supplies nitrogen to the succeeding wheat crop.

Methodology

Field experiments were conducted for ten years (2005-06 to 2014-15) at the research farm of CCS Haryana Agricultural University, Hisar India in mungbean-wheat (*desi*) cropping system. The soil is a typic *Haplustalf*, loosely aggregated; and sandy loam in texture. The surface (0-15 cm) soil has 58.8 per cent sand, 17.3 silt and 23.9 per cent clay. The initial organic carbon content, pH and EC of the surface (0-15 cm) soil were 0.48 per cent, 7.7 and 0.12 dS m^{-1}, respectively. Available nitrogen (186 kg ha^{-1}), phosphorus (25.6 kg ha^{-1}) and potassium (290 kg ha^{-1}) graded as low, medium and high, respectively. The field was well levelled and had no salinity or drainage problems.

The climate of Hisar is semi-arid. The temperature in summers rises and had been reported to be as high as 48°C and in winters the temperature goes as low as 0 to -2°C. Hot winds blow during summer (May-June) and cold wave is experienced in winter (Dec.-Jan.). In Hisar, about 85 per cent rain is received from south-western monsoon from July to September. In general weather aberrations are too frequent in the area. Total rainfall was 560.9, 508.1, 353.6, 495.5, 420.1, 915.9 416.8, 603.9, 809.7 and 619.9 mm during 2005-06 to 2014-15, respectively.

The experiment was initiated in *Kharif* 2005, with the sowing of mung bean and reported up to *Rabi* 2014-15 with the harvesting of wheat. The experiment was laid out in 7 strips each of 60 m length and 5.4 m breadth.

Vermi-compost, FYM, neem cake, rock phosphate, N and P bio-fertilizers (rhizobium for mung bean and azotobacter in wheat as source of nitrogen and phosphate solublising bacteria as source of phosphorus) were the organic sources of nutrients used in both the crops. Urea, DAP, MOP and $ZnSO_4$ in treatment $T_1$and T_2 were the chemical sources of nutrients used in mungbean and wheat (*desi*). Mung bean variety Asha/Satya and wheat variety C-306 were sown in July and October/ November, respectively.

The organic source namely FYM, vermi-compost and neem-cake were applied to both crops. The required quantity of these sources was applied in the strip as per treatment and was incorporated in the soil before sowing of crops.

Wheat equivalent yield (WEY) was calculated on the basis of yield and price of crops. Economics was calculated on the basis of prevailing market rates/ university rates/minimum support price for inputs and produce for each year. The recommended dose of nutrients for mung bean is 20 kg nitrogen and 40 kg phosphorus per ha and for wheat(*desi*) is 60 kg nitrogen, 30 kg phosphorus, 15 kg potash and 25 kg zinc sulphate per ha.

Treatment Details

$T_{1:}$ 50 per cent recommended NPK + 50 per cent N (FYM); T_2: $^1/_3{}^{rd}$ N each through FYM, vermi-compost, neem cake; T_3: T_2 + Intercropping (*Kharif*: Pearl millet, *Rabi*: Mustard); T_4: T_2 + Agro. practices for weed and pest control; T_5: 50 per cent N FYM + bio-fertilizer N + rock phosphate + PSB; T_6: T_2 + N+P bio-fertilizer; T_7: Recommended fertilizer.

Observations on grain yield of mungbean and wheat, quality parameters of wheat (*desi*) were estimated as per standard protocol. Sedimentation values were analyzed by dodecyl sulphate (SDS) test as suggested by Axford *at el, 1979*. Hectoliter weight was estimated through test weight instrument developed at DWR Karnal. Return over variable cost/net returns were computed keeping in account input and output cost as per rates fixed/approved by government, NPK content and uptake by both the crops, NPK availability in soil and EC, pH and OC of soil before sowing and after harvesting of each crop was analyzed using standard methods.

Bumper Crop of Organic Wheat.

1/3 N each through
FYM, Vermi-compost
and Neem Cake

Results and Discussion

Crop Productivity

Initial Period

During initial year of study, the yield of mungbean in *Kharif* season ranged between 732 to 1195 kg ha^{-1}. Highest yield of 1195 kg ha^{-1} was recorded with the application of chemical fertilizers (T_7) which was followed by application of 50 per cent recommended NPK + 50 per cent N through FYM with yield of 1120 kg ha^{-1}. Integration of organic and inorganic sources of nutrients increases the yield as compared to organics alone. In this study integration of chemical and organic source (T_1) performed better as compared to organic sources (T_2, T_3, T_4, T_5 and T_6) alone. Among organic sources of nutrients, 1/3 N each through FYM, vermicompost and neemcake (T_2) yielded 816 kg ha^{-1} and it was 928 kg ha^{-1} when seed was treated with rhizobium culture and phosphate solublizing bacteria (PSB) before sowing (T_6). The increase in yield was 112 kg ha^{-1} when the seed was treated with bio-fertilizers (Table 1).

During initial year of study the yield of wheat was higher (2870 kg ha^{-1}) with the application of chemical fertilizers (T_7) and was followed by treatment where 50 per cent recommended NPK+50 per cent N through FYM (T_1) was applied with yield

Table 1: Effect of Nutrient Management Package on Crop Yield (kg ha^{-1}) in Mungbean - Wheat (*Desi*) Cropping System in different Phases of Crop Cycles

Treatments	*Initial yield level*			*Mean Yield upto Conversion Period (1st -3rd Crop Cycles)*			*Mean Yield After Conversion Period (4th-10th crop cycles)*			*Overall Mean Yield (10th crop cycles)*		
	Kharif	*Rabi*	*SEY*	*Kharif*	*Rabi*	*SEY*	*Kharif*	*Rabi*	*SEY*	*Kharif*	*Rabi*	*SEY*
T_1	1120	2330	3450	1140	2422	3798	1135	2526	6329	1132	2426	4526
T_2	816	1875	2740	862	1971	3030	898	2154	5186	859	2000	3652
T_3	786	1800	2586	772	1848	2962	713	1913	4829	757	1854	3459
T_4	851	1850	2701	860	1945	2983	945	2182	5372	885	1992	3685
T_5	732	1685	2417	692	1711	2541	565	1481	3384	663	1626	2781
T_6	928	1970	2898	929	2067	3186	857	2199	5095	905	2079	3726
T_7	1195	2870	4065	1208	3007	4463	1224	3024	7162	1209	2967	5230
SEM (±)				14	22	46	25	32	137	72	184	136

of 2330 kg ha^{-1}. Yield of wheat increased with the integration of chemical fertilizers with FYM (T_1) as compared to organic sources of nutrients (T_2, T_3, T_4 and T_5).

System equivalent yield of mungbean-wheat cropping system ranged between 2417 to 4065 kg ha^{-1} being highest (4065 kg ha^{-1}) with chemical fertilizers (T_7) and lowest (2417 kg ha^{-1}) with the application of 50 per cent N through FYM + bio-fertilizers + rock phosphate.

Up to Conversion Period

The yield level up to initial three crop cycles followed the pattern as indicated during initial year of experimentation. The highest yield of mungbean was recorded 1208 kg ha^{-1} with the application of chemical fertilizers (T_7) and it was followed by integration of chemical and organic sources of nutrients (T_1) with the yield of 1140 kg ha^{-1}.

The wheat yield of 3007 kg ha^{-1} was recorded up to the period of conversion (1st-3rd crop cycle) with the application of chemical fertilizers (T_7) and it was followed by application of 50 per cent recommended NPK + 50 per cent N through FYM (T_1) with the yield of 2422 kg ha^{-1}.

System equivalent yield of 4463 kg ha^{-1} was recorded with the application of 100 per cent recommended nutrients through chemical fertilizers (T_7) and it was followed by 3798 kg ha^{-1} with 50 per cent recommended NPK + 50 per cent N through FYM. Among organic sources of nutrients the system yield ranged between 2541 to 3186 kg ha^{-1}. Singh *et al.* (2008) has also reported higher equivalent yield in mung bean-wheat cropping system because mungbean as preceding crop improves physical, chemical and microbial properties of soil.

After Conversion Period

The yield of mungbean followed the same trend during conversion period (4th to 10th crop cycles) as followed by up to conversion period (1st-3rd crop cycle). The highest yield of mungbean 1224 and wheat 3024 kg ha^{-1} was recorded with the application of chemical fertilizers (T_7) and followed by the treatment where 50 per cent N recommended NPK + 50 per cent N was applied (T_1) with the yield of 1135 kg ha^{-1} and 2625 kg ha^{-1} in respect of mungbean and wheat. System yield of 7162 kg ha^{-1} was recorded highest after the completion of conversion period with the application of chemical fertilizers. The yield through application of organic sources (T_2 to T_6) ranged between 3384 to 5372 kg ha^{-1}, whereas, integration of chemical and organic sources produced yield of 6329 kg ha^{-1} (Table 1). The higher yield through integrated use of chemical and organic sources might be due to efficient and greater partitioning of metabolites and adequate translocation of photosynthates and nutrients to developing reproductive structure. Behra *et al.* (2007) also reported the similar result.

The overall mean of 10 crop cycles indicate the same trend as shown by individual crop and system as it was during conversion period to period after conversion.

Correlation Study

In general, the system (mungbean-wheat) yield was highly correlated (Fig. 1) with the sources of nutrients irrespective of level of application except in T_5. Slightly lower value of coefficient of regression (0.84) was obtained when 50 per cent N through FYM+N and P bio-fertilizers + rock phosphate was applied. It is also clear from Table 2 that system yield responded positively when nitrogen was applied through a combination of inorganic and organic source of nutrients.

Table 2: Yield Trend Equation for System Yield in Mungbean - Wheat (*Desi*) Cropping System during 10 Years of Investigation (2005-2014) "Best Fit Polynomial Equation"

Treatments	*System Yield*	R^2
T_1	$Y = 21.75X^2 + 337.48X + 3077.8$	0.9885
T_2	$Y = 26.999X^2 + 215.46X + 2508.3$	0.9958
T_3	$Y = 58.076X^2 - 170.92X + 2998.7$	0.9468
T_4	$Y = 30.732X^2 + 198.18X + 2489.7$	0.9924
T_5	$Y = 13.886X^2 + 60.686X + 2407.5$	0.8444
T_6	$Y = 5.6385X^2 + 387.21X + 2420.9$	0.9912
T_7	$Y = 24.267X^2+375.58X+3665.4$	0.9861

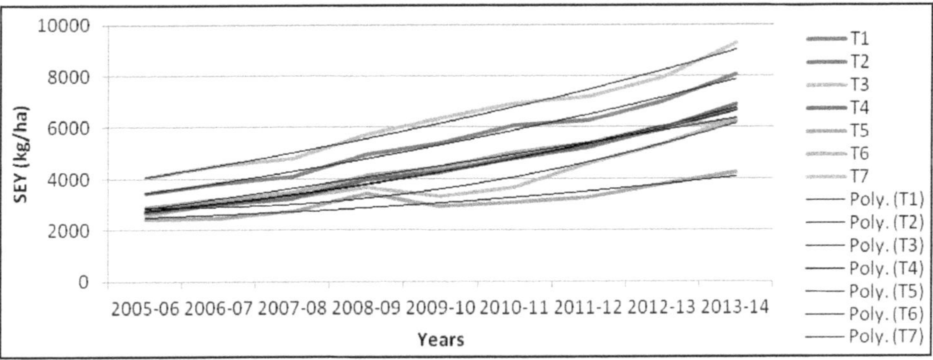

Figure 1: Correlation Study in Mungbean-Wheat Cropping System with different Sources of Nutrients.

Return over Variable Cost/Net Returns

The return over variable cost/net income (mean of 10 year) among various treatments varied between Rs. 13743 and Rs. 62612 ha^{-1} (Table 3). Highest net income of Rs. 62612 ha^{-1} was recorded with the application of recommended chemical fertilizers (T_7). The next best treatment with a net income of Rs. 50038 ha^{-1} was where 50 per cent recommended dose was applied through chemical fertilizers and remaining 50 per cent N was applied through FYM. Net income was higher with the application of chemical fertilizers followed by integration of chemical and organic sources of nutrient as compared to organic sources of nutrients. Bhattacharyya *et al.* (2008) also reported that integration of organic and inorganic sources of nutrients

Table 3: Effect of Nutrient Management Packages on Return over Variable Cost/Net Return (Rs./ha) in Mung Bean-Wheat Cropping System in different Phases of Crop Cycle

Treatment	Year										
	2005- 06	2006-07	2007-08	2008-09	2009-10	2010-11	2011-12	2012-13	2013-14	2014-15	Mean
T_1	26402	35710	38640	38875	26931	48460	52882	65938	74505	92035	50038
T_2	10111	19141	21708	9361	4943	12390	17630	26902	32141	45746	20007
T_3	6932	15516	18470	7744	4170	3695	10114	17538	24522	39809	14851
T_4	9660	18153	21010	7124	5320	11920	16800	27475	33215	92723	24340
T_5	9413	12997	14894	10744	4285	8633	11550	18620	20234	26061	13743
T_6	13391	21545	23516	12445	5542	15410	19277	29267	26190	38004	20459
T_7	35781	44760	46845	50638	39947	61777	68058	81674	88018	108618	62612

gave higher monetary returns. The return over variable cost/net returns were influenced by crop yield and minimum support price/sale price of produce.

Crop Quality

Hectolitre weight, sedimentation value, gluten content and protein content were slightly affected by source of nutrients (Table 4). Hectolitre weight (79.78 g) was highest where recommended dose of nutrients were applied through chemical fertilizers. Sedimentation value in organic treatments ranged between 34 and 36. Per cent gluten content among organic treatments was between 21.40 to 31.29 and per cent protein content was between 10.6 and 11.6. Numerically highest protein content (11.7 per cent) was analysed in treatment T_7 where recommended fertilizer dose was applied to both crops.

Table 4: Effect of Nutrient Management Packages on Quality Parameters of Wheat (mean after conversion period)

Treatments	Hectoliter Weight (hl)	Sedimentation Value (Per cent)	Gluten Content (Per cent)	Protein Content (Per cent)
T_1	79.74	34	30.15	11.5
T_2	78.43	35	31.17	11.3
T_3	78.09	35	27.40	11.3
T_4	78.87	35	28.46	11.6
T_5	78.90	34	31.29	11.6
T_6	78.40	36	29.91	10.6
T_7	79.78	32	29.60	11.7

Nutrients Uptake

Mean of initial three years (2005-06 to 2007-08) N uptake during *Kharif* season varied between 66.1 and 123.0 kg ha^{-1} (Table 5) among different treatments. While the mean of next six years of experimentation varied between 58.5 and 142.3 kg ha^{-1}. The overall mean variation (2005-06 to 2014-15) among different treatments was between 62.3 and 132.6 kg ha^{-1} being minimum in T_5 where application of 50 per cent N through FYM along with bio-fertilizer N, rock phosphate and PSB bio-fertilizer was made and maximum in the treatment T_7 where 100 per cent recommended dose of fertilizers was applied, respectively. Integrated use of fertilizers *i.e.* in treatment T_1 where 50 per cent recommended dose of fertilizers were applied along with 50 per cent N which was added through FYM resulted in N uptake of 123.1 kg ha^{-1}. While after the harvest of *Rabi* crop *i.e.* wheat (*desi*) (Table 6) the initial three years (2005-06 to 2007-08) mean N varied between 52.1 kg ha^{-1} in treatment T_5 where application of 50 per cent N through FYM along with bio-fertilizer N, rock phosphate and PSB bio-fertilizer was applied and 98.1 kg ha^{-1} in T_7 where 100 per cent recommended dose of fertilizers was applied. Similar trend was followed with 39.7 kg ha^{-1} mean N uptake in T_5 and 88.7 kg ha^{-1} in treatment T_7. Overall mean N uptake varied between 41.4 and 89.4 kg ha^{-1}, being minimum and maximum in treatments T_5 and

T_7, respectively. On the whole trend with respect to mean N uptake both during *Kharif* and *Rabi* was same but of different magnitude.

Table 5: Mean *Kharif* Nitrogen Uptake (kg/ha) of 3-years (2005-06 to 2007-08), 7-years (2008-09 to 2014-15) and Overall (2005-06 to 2014-15) among different Treatments in Organic Farming Package for Mung Bean – Wheat (*Desi*) Cropping System

		Mean Kharif Nitrogen Uptake		
		2005-06 to 2007-08	*2008-09 to 2014-15*	*2005-06 to 2014-15*
T_1	50 per cent Rec. NPK + 50 per cent N (FYM)	122.2	124.0	123.1
T_2	N each through FYM, V.C., Neem cake	90.4	99.0	94.7
T_3	T_2 + Intercropping (*Kharif* : Pearlmillet, *Rabi* : Mustard)	76.5	132.1	104.3
T_4	T_2 + Agro. Practices for weed and pest control	89.9	104.9	97.4
T_5	50 per cent N FYM + N bio-fertilizer + rock phosphate + PSB	66.1	58.5	62.3
T_6	T_2 + N+P bio-fertilizer	99.9	95.3	97.6
T_7	Recommended fertilizer	123.0	142.3	132.6

Table 6: Mean *Rabi* Nitrogen Uptake (kg/ha) of 3-years (2005-06 to 2007-08), 7-years (2008-09 to 2014-15) and Overall (2005-06 to 2014-15) among different Treatments in Organic Farming Package for Mung Bean – Wheat (*Desi*) Cropping System

Treatment		*Mean Rabi Nitrogen Uptake*		
		2005-06 to 2007-08	*2008-09 to 2014-15*	*2005-06 to 2014-15*
T_1	50 per cent Rec. NPK + 50 per cent N (FYM)	70.9	72.2	71.2
T_2	N each through FYM, V.C., Neem cake	62.8	60.5	59.7
T_3	T_2 + Intercropping (*Kharif* : Pearlmillet, *Rabi* : Mustard)	65.0	60.6	60.4
T_4	T_2 + Agro. Practices for weed and pest control	64.2	61.3	60.6
T_5	50 per cent N FYM + N bio-fertilizer + rock phosphate + PSB	52.1	39.7	41.4
T_6	T_2 + N+P bio-fertilizer	61.7	61.9	60.8
T_7	Recommended fertilizer	98.1	88.7	89.4

Table 7: Mean *Kharif* Phosphorus Uptake (kg/ha) of 3-years (2005-06 to 2007-08), 7-years (2008-09 to 2014-15) and Overall (2005-06 to 2014-15) among different Treatments in Organic Farming Package for Mung Bean – Wheat (*Desi*) Cropping System

Treatment		*Mean Kharif Phosphorus Uptake*		
		2005-06 to 2007-08	*2008-09 to 2014-15*	*2005-06 to 2014-15*
T_1	50 per cent Rec. NPK + 50 per cent N (FYM)	18.6	21.1	21.9
T_2	N each through FYM, V.C., Neem cake	13.2	16.8	17.1
T_3	T_2 + Intercropping (*Kharif* : Pearlmillet, *Rabi* : Mustard)	12.6	14.5	14.5
T_4	T_2 + Agro. Practices for weed and pest control	13.4	18.5	18.3
T_5	50 per cent N FYM + N bio-fertilizer + rock phosphate + PSB	9.7	11.1	11.7
T_6	T_2 + N+P bio-fertilizer	15.0	16.5	17.6
T_7	Recommended fertilizer	17.7	25.0	25.1

Table 8: Mean *Rabi* Phosphorus Uptake (kg/ha) of 3-years (2005-06 to 2007-08), 7-years (2008-09 to 2014-15) and Overall (2005-06 to 2014-15) among different Treatments in Organic Farming Package for Mung Bean – Wheat (*Desi*) Cropping System

Treatment		*Mean Rabi Phosphorus Uptake*		
		2005-06 to 2007-08	*2008-09 to 2014-15*	*2005-06 to 2014-15*
T_1	50 per cent Rec. NPK + 50 per cent N (FYM)	13.3	14.2	14.0
T_2	N each through FYM, V.C., Neem cake	11.2	11.7	11.4
T_3	T_2 + Intercropping (*Kharif* : Pearlmillet, *Rabi* : Mustard)	11.6	12.3	12.0
T_4	T_2 + Agro. Practices for weed and pest control	11.9	12.1	12.0
T_5	50 per cent N FYM + N bio-fertilizer + rock phosphate + PSB	9.8	8.7	8.8
T_6	T_2 + N+P bio-fertilizer	12.3	12.5	12.2
T_7	Recommended fertilizer	18.4	18.1	18.1

Table 9: Mean *Kharif* Potash Uptake (kg/ha) of 3-years (2005-06 to 2007-08), 7-years (2008-09 to 2014-15) and Overall (2005-06 to 2014-15) among different Treatments in Organic Farming Package for Mung Bean – Wheat (*Desi*) Cropping System

Treatment		Mean Kharif Potash Uptake		
		2005-06 to 2007-08	2008-09 to 2014-15	2005-06 to 2014-15
T_1	50 per cent Rec. NPK + 50 per cent N (FYM)	101.4	70.7	80.6
T_2	N each through FYM, V.C., Neem cake	82.5	55.5	63.4
T_3	T_2 + Intercropping (*Kharif* : Pearlmillet, *Rabi* : Mustard)	82.5	57.1	62.8
T_4	T_2 + Agro. Practices for weed and pest control	84.1	59.1	66.0
T_5	50 per cent N FYM + N bio-fertilizer + rock phosphate + PSB	65.6	34.2	42.5
T_6	T_2 + N+P bio-fertilizer	90.2	53.3	64.9
T_7	Recommended fertilizer	107.0	79.7	89.7

Table 10: Mean *Rabi* Potash Uptake (kg/ha) of 3-years (2005-06 to 2007-08), 7-years (2008-09 to 2014-15) and Overall (2005-06 to 2014-15) among different Treatments in Organic Farming Package for Mung Bean – Wheat (*Desi*) Cropping System

Treatment		Mean Rabi Potash Uptake		
		2005-06 to 2007-08	2008-09 to 2013-14	2005-06 to 2013-14
T_1	50 per cent Rec. NPK + 50 per cent N (FYM)	135.9	134.0	134.3
T_2	N each through FYM, V.C., Neem cake	114.4	112.5	109.0
T_3	T_2 + Intercropping (*Kharif* : Pearlmillet, *Rabi* : Mustard)	109.9	115.5	112.0
T_4	T_2 + Agro. Practices for weed and pest control	112.4	117.2	114.5
T_5	50 per cent N FYM + N bio-fertilizer + rock phosphate + PSB	92.4	78.8	81.0
T_6	T_2 + N+P bio-fertilizer	115.6	123.8	119.6
T_7	Recommended fertilizer	180.6	160.9	164.1

Mean P uptake during *Kharif* ranged between 9.7 and 18.6 kg ha^{-1} in treatment T_5 and maximum in treatment T_1 during the initial three years of experiment. The mean of next six years experiment varied between 11.1 and 25.0 kg ha^{-1} being minimum in T_5 where application of 50 per cent N through FYM along with bio-fertilizer N, rock phosphate and PSB bio-fertilizer was made and maximum in T_7 where full dose of recommended fertilizers were applied. Overall mean (2005-06 to 2014-15) P uptake during *Kharif* varied between 11.7 and 25.1 kg ha^{-1} in treatments T_5 and T_7, respectively (Table 8). Mean P uptake of after *Rabi* harvest (Table 9) during initial three years of experiments showed variation between 9.8 and 18.4 kg ha^{-1}. While the mean of further six years of experiments showed variation between 8.7 and 18.1 kg ha^{-1} with respect to P uptake. The overall mean (2005-06 to 2014-15) P uptake varied between 8.8 and 18.1 kg ha^{-1}, with minimum and maximum uptake as followed earlier by the means of three and six years of experiments.

Mean of initial three years (2005-06 to 2007-08) potash uptake during *Kharif* varied between 65.6 kg/ha in treatment T_5 and 107.0 kg/ha in treatment T_7 (Table 9). However the mean of further six years *i.e.* 2008-09 to 2014-15 ranged between 34.2 and 79.7 kg/ha; being minimum and maximum in treatment T_5 and T_7, respectively. The overall variation (2005-06 to 2014-15) among different treatments varied between 42.5 and 89.7 kg/ha; being minimum in treatment T_5 and maximum in treatment T_7. Integrated use of fertilizers (T_1) resulted in overall mean potash uptake of 80.6 kg/ha. In treatment T_2 where 1/3 N was added each through FYM, vermin-compost and neem cake resulted in overall mean potash uptake of 63.4 kg/ha. While after the harvest of *Rabi* crop *i.e.* wheat (*desi*), the initial three years (2005-06 to 2007-08) mean N uptake varied between 92.4 kg/ha in treatment T_5 and 180.6 kg/ha in treatment T_7 (Table 10). Mean potash uptake of further six years (2008-09 to 2014-15) resulted in uptake of 134.0 kg/ha in treatment T_1 where integrated use of organic manure and chemical fertilizers was made. The range among different treatments with respect to mean potash uptake was between 78.8 and 160.9 kg/ha; being minimum in T_5 and maximum in T_7, respectively. But the overall (2005-06 to 2014-15) mean potash uptake ranged between 81.0 kg/ha in treatment T_5 and 164.1 kg/ha in treatment T_7. Treatment T_1 resulted in mean potash uptake of 134.3 kg/ha. In T_2 treatment where 1/3 N was applied each through FYM, vermi-compost and neem cake the value was 109.0 kg/ha.

Table 11: Regression Equations on *Kharif* Nitrogen Uptake (kg ha^{-1}) Changes under Organic Nutrient Management Packages in Mungbean-Wheat (*Desi*) Cropping System

Treatments	*Regression Equations*	*R^2 Values*
T_1	$Y = -3.7088X^2 + 37.976X + 65.198$	0.39
T_2	$Y = -3.0129X^2 + 32.534X + 39.222$	0.42
T_3	$Y = -1.4808X^2 + 15.872X + 52.595$	0.22
T_4	$Y = -3.2284X^2 + 34.653X + 37.358$	0.48
T_5	$Y = -2.0952X^2 + 19.127X + 40.612$	0.31
T_6	$Y = -4.1244X^2 + 41.668X + 37.251$	0.46
T_7	$Y = -4.6169X^2 + 50.964X + 43.844$	0.62

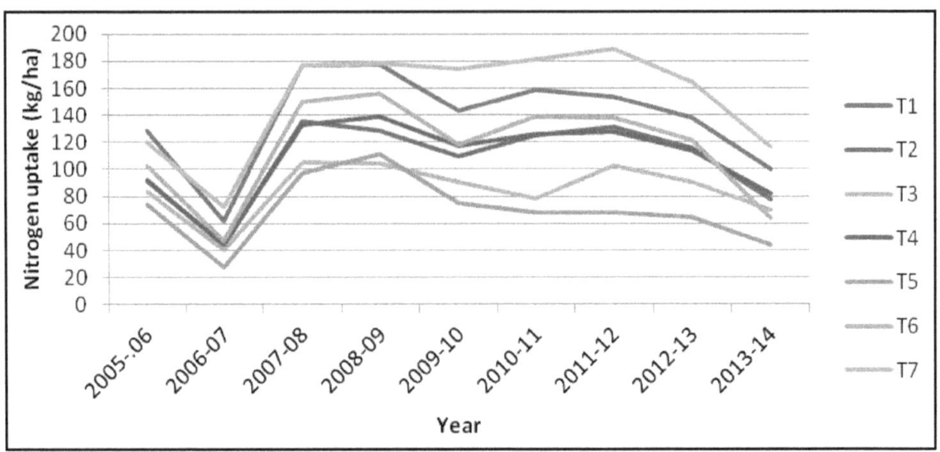

Figure 2: Regression Equations on *Kharif* Nitrogen Uptake (kg ha^{-1}) Changes Under Organic Nutrient Management Treatments.

Table 12: Regression Equations on *Rabi* Nitrogen Uptake (kg ha^{-1}) Changes under Organic Nutrient Management Package in Mungbean-Wheat (*Desi*) Cropping System

Treatments	*Regression Equations*	*R^2 Values*
T_1	$Y = -0.1463X^2 + 1.2599X + 68.612$	0.07
T_2	$Y = 0.6042X^2 - 6.8073X + 73.741$	0.71
T_3	$Y = 0.4709X^2 - 6.117X + 75.841$	0.37
T_4	$Y = 0.7809X^2 - 8.923X + 79.734$	0.64
T_5	$Y = 0.6564X^2 - 9.1921X + 68.292$	0.84
T_6	$Y = 0.2315X^2 - 2.6016X + 65.294$	0.15
T_7	$Y = 0.612X^2 - 8.7767X + 114.58$	0.80

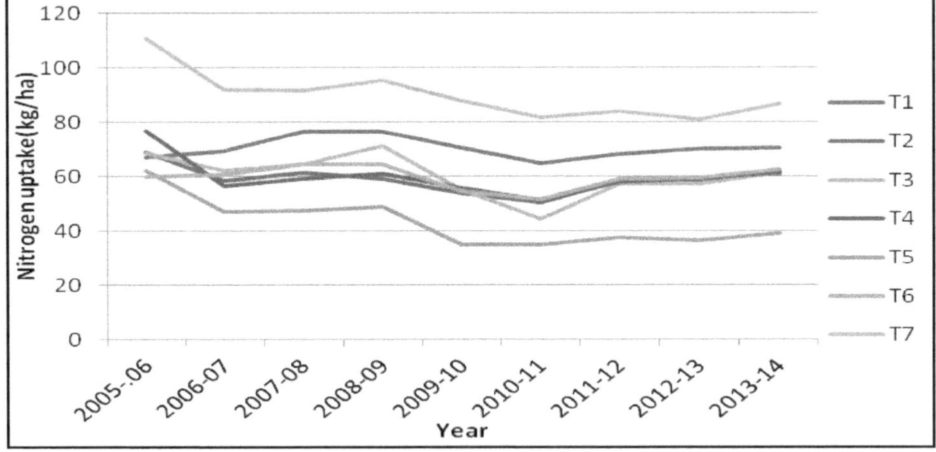

Figure 3: Regression Equations on *Rabi* Nitrogen Uptake (kg ha^{-1}) Changes Under Organic Nutrient Management Treatments.

Table 13: Regression Equations on *Kharif* Phosphorus Uptake (kg ha^{-1}) Changes under Organic Nutrient Management Package in Mungbean-Wheat (*Desi*) Cropping System

Treatments	*Regression Equations*	*R^2 Values*
T_1	$Y = -0.8904X^2 + 9.6669X + 2.5379$	0.70
T_2	$Y = -0.7279X^2 + 8.2092X - 0.6286$	0.74
T_3	$Y = -0.444X^2 + 4.8424X + 4.3171$	0.44
T_4	$Y = -0.7392X^2 + 8.28X - 0.0264$	0.80
T_5	$Y = -0.5946X^2 + 6.1508X + 0.2521$	0.59
T_6	$Y = -0.9442X^2 + 10.001X - 1.359$	0.75
T_7	$Y = -1.1277X^2 + 12.862X - 3.3998$	0.86

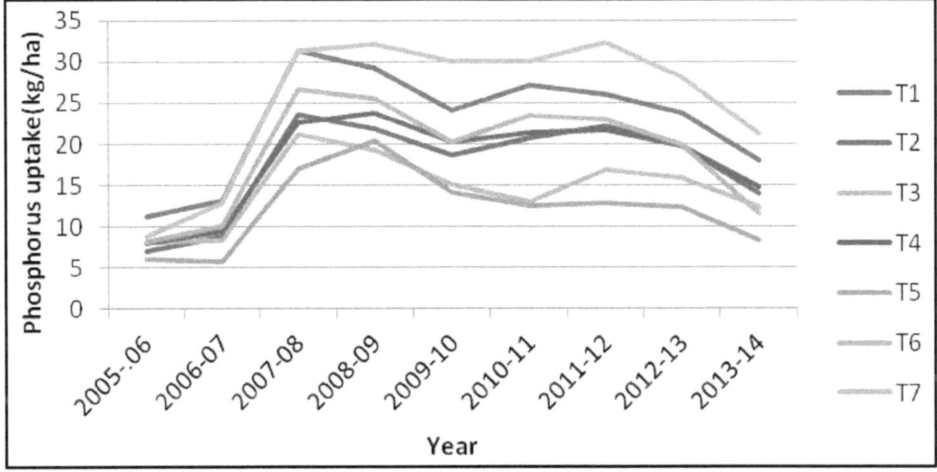

Figure 4: Regression Equations on *Kharif* Phosphorus Uptake (kg ha^{-1}) Changes under Organic Nutrient Management Package.

Table 14: Regression Equations on *Rabi* Phosphorus Uptake (kg ha^{-1}) Changes under Organic Nutrient Management Package in Mungbean-Wheat (*Desi*) Cropping System

Treatments	*Regression Equations*	*R^2 Values*
T_1	$Y = -0.1682X^2 + 1.7734X + 10.247$	0.37
T_2	$Y = 0.0033X^2 + 0.0495X + 10.784$	0.07
T_3	$Y = 0.0047X^2 + 0.0823X + 11.265$	0.03
T_4	$Y = -0.0145X^2 + 0.1448X + 11.523$	0.01
T_5	$Y = 0.0633X^2 - 0.8031X + 10.994$	0.51
T_6	$Y = -0.0052X^2 + 0.0004X + 12.107$	0.02
T_7	$Y = -0.1598X^2 + 1.4206X + 16.027$	0.24

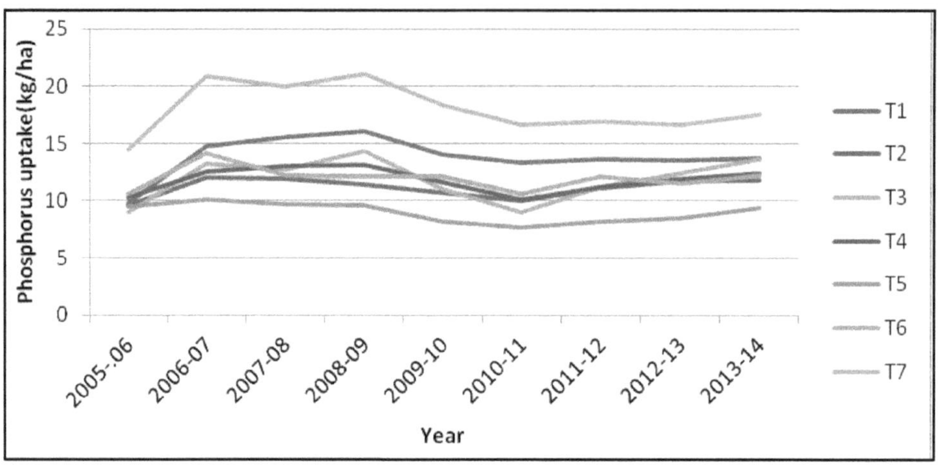

Figure 5: Regression Equations on *Rabi* Phosphorus Uptake (kg ha^{-1}) Changes under Organic Nutrient Management Package.

Table 15: Regression Equations on *Khrif* Potash Uptake (kg ha^{-1}) Changes under Organic Nutrient Management Package in Mungbean-Wheat (*Desi*) Cropping System

Treatments	*Regression Equations*	*R^2 Values*
T_1	$Y = -0.8684X^2 + 4.1472X + 97.413$	0.76
T_2	$Y = -0.1052X^2 - 2.7614X + 88.471$	0.62
T_3	$Y = 0.2944X^2 - 6.9637X + 94.086$	0.76
T_4	$Y = -0.1279X^2 - 2.7388X + 90.713$	0.74
T_5	$Y = -0.086X^2 - 4.4959X + 75.939$	0.91
T_6	$Y = -0.6693X^2 + 1.1641X + 91.922$	0.74
T_7	$Y = -0.8577X^2 + 5.0499X + 101.68$	0.63

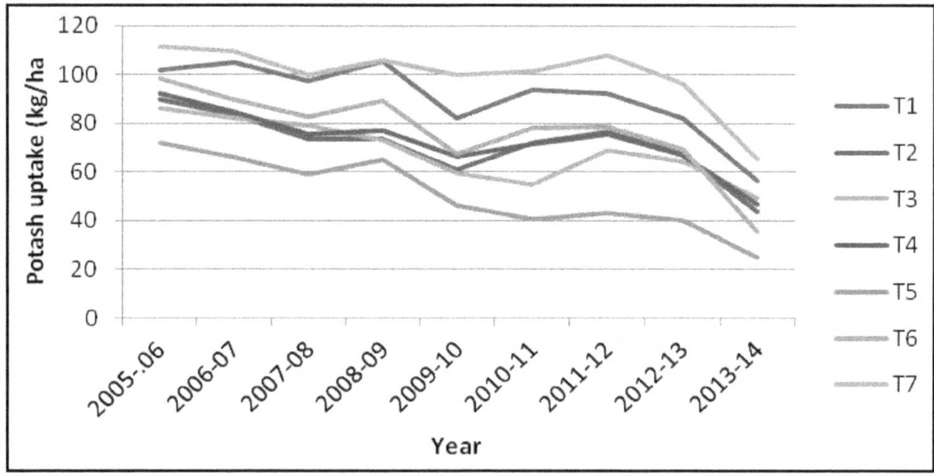

Figure 6: Regression Equations on *Kharif* Potash Uptake (kg ha^{-1}) Changes under Organic Nutrient Management Package.

Table 16: Regression Equations on *Rabi* Potash Uptake (kg ha^{-1}) Changes under Organic Nutrient Management Package in Mungbean-Wheat (*Desi*) Cropping System

Treatments	*Regression Equations*	*R^2 Values*
T_1	$Y = -0.6869X^2 + 5.7876X + 127.41$	0.22
T_2	$Y = -0.0042X^2 - 3.3228X + 122.35$	0.17
T_3	$Y = 1.2043X^2 - 11.193X + 129.22$	0.39
T_4	$Y = 0.6668X^2 - 5.7353X + 120.58$	0.17
T_5	$Y = 0.6914X^2 - 8.3914X + 101.34$	0.38
T_6	$Y = 0.3775X^2 - 3.3549X + 120.8$	0.06
T_7	$Y = 0.4063X^2 - 7.9408X + 189.27$	0.45

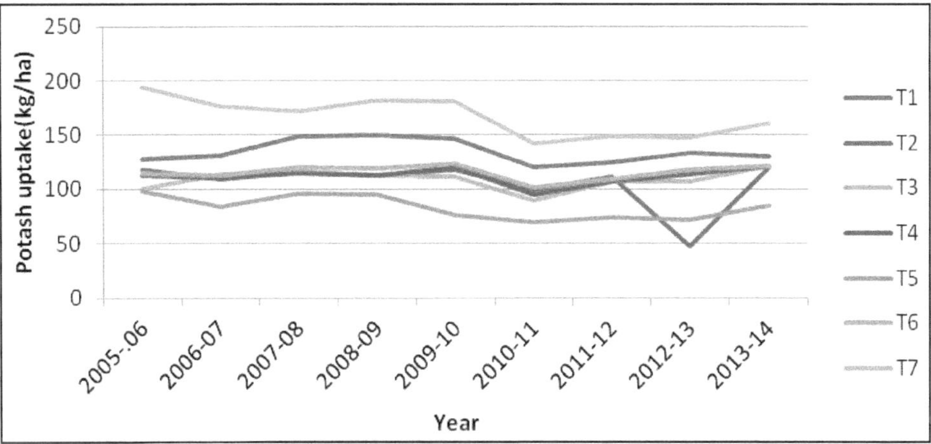

Figure 7: Regression Equations on *Rabi* Potash Uptake (kg ha^{-1}) Changes under Organic Nutrient Management Package.

Soil Fertility Status

Initial soil fertility status has been given in Table 18. Mean of initial three years (2005-06 to 2007-08) showed marked variation in the range of 0.44 and 0.51 g/kg, whereas, mean of six years *i.e.* 2008-09 to 2014-15 showed variation between 0.47 to and 0.51 g/kg among different treatments (Table 18). The overall variation in organic carbon among different treatments *i.e.* mean of 10 years of experimentation ranged between 0.46 and 0.51 g/kg. Among all the means of three, six and nine years, the trend with respect to organic carbon in soil showed similar trend *i.e.* being maximum in T_7 where 100 per cent recommended dose of fertilizers were applied and minimum in T_5 where application of 50 per cent N through FYM along with bio-fertilizer N, rock phosphate and PSB bio-fertilizer was made. Lowest organic carbon was observed in treatment T_5. This might be due to less solubility of organic sources in soil thereby resulting in less availability of nutrients in soil which ultimately led to less root proliferation and thereby less bio-mass and ultimately less organic matter production and thus influenced organic carbon status. On comparing mean

Table 17: Changes in Total Nutrient Uptake (kg ha^{-1}) under Various Nutrient Management Package in Mungbean-Wheat (*Desi*) Cropping System

Treatments	*Initial Nutrient Uptake (kg ha^{-1}) i.e. at End of 1st Crop Cycle*				*Average (kg ha^{-1}) i.e from First Crop Cycle to Last Crop Cycle*				*per cent Changes over Initial (Kast year vs. First year)*				*Changes Over T_7 Average*			
	N	*p*	*K*	*Total*	*N*	*p*	*K*	*Total*	*N*	*p*	*K*	*Total*	*N*	*p*	*K*	*Total*
T_1	67	9.8	127.9	204.7	70.3	13.8	134.6	218.7	5.22	39.80	1.64	46.66	19.8	4.3	32.7	56.8
T_2	68.9	9.6	117.9	196.4	58.8	11.1	105.6	175.5	-11.76	22.92	1.53	12.69	31.3	7	61.7	100
T_3	68.6	9	100.8	178.4	60.2	11.8	108.6	180.6	-10.64	51.11	20.04	60.51	29.9	6.3	58.7	94.9
T_4	77	10.3	112	199.3	59.8	11.8	111.8	183.4	-20.52	20.39	7.77	7.64	30.3	6.3	55.5	92.1
T_5	62	9.5	97.9	169.4	43.1	9.0	83.1	135.2	-37.26	-1.05	-13.38	-51.69	47	9.1	84.2	140.3
T_6	60	10.6	115.1	185.7	59.6	11.9	115.3	186.8	3.83	15.09	5.04	23.96	30.5	6.2	52	88.7
T_7	110.9	14.4	194.1	125.3	90.1	18.1	167.3	275.5	-22.00	22.22	-16.85	-16.63				

Table 18: Mean Organic Carbon (g/kg) of 3-years (2005-06 to 2007-08), 7-years (2008-09 to 2014-15) and Overall (2005-06 to 2014-15) among different Treatments in Organic Farming Package for Mung Bean-Wheat (*Desi*) Cropping System

Treatment		*Mean Organic Carbon*		
		2005-06 to 2007-08	*2008-09 to 2014-15*	*2005-06 to 2014-15*
T_1	50 per cent Rec. NPK + 50 per cent N (FYM)	0.48	0.49	0.49
T_2	N each through FYM, V.C., Neem cake	0.49	0.49	0.49
T_3	T_2 + Intercropping (*Kharif* : Pearlmillet, *Rabi* : Mustard)	0.48	0.50	0.49
T_4	T_2 + Agro. Practices for weed and pest control	0.50	0.50	0.50
T_5	50 per cent N FYM + N bio-fertilizer + rock phosphate + PSB	0.44	0.47	0.46
T_6	T_2 + N+P bio-fertilizer	0.50	0.48	0.49
T_7	Recommended fertilizer	0.51	0.51	0.51

Table 19: Mean Available Nitrogen (kg/ha) of 3-years (2005-06 to 2007-08), 7-years (2008-09 to 2014-15) and Overall (2005-06 to 2014-15) among different Treatments in Organic Farming Package for Mung Bean-Wheat (*Desi*) Cropping System

Treatment		*Mean Available Nitrogen*		
		2005-06 to 2007-08	*2008-09 to 2014-15*	*2005-06 to 2014-15*
T_1	50 per cent Rec. NPK + 50 per cent N (FYM)	173.6	166.35	166.6
T_2	N each through FYM, V.C., Neem cake	165.2	158.65	159.4
T_3	T_2 + Intercropping (*Kharif* : Pearlmillet, *Rabi* : Mustard)	168.5	162.15	162.2
T_4	T_2 + Agro. Practices for weed and pest control	156.9	162.50	160.7
T_5	50 per cent N FYM + N bio-fertilizer + rock phosphate + PSB	146.5	155.05	152.6
T_6	T_2 + N+P bio-fertilizer	160.5	163.80	162.3
T_7	Recommended fertilizer	184.3	168.45	170.3

organic carbon of three years with that of overall (2005-06 to 2014-15) treatment T_5 showed increase of 0.20 g/kg and T_7 showed no decrease. Organic carbon being the most important indicator of soil fertility indicates that organic farming is capable of sustaining higher crop productivity and improving soil quality and productivity by manipulating the soil properties on long term basis. It was reported that organic and low-input farming practices after four years led to an increase in organic carbon, soluble P, exchangeable K, and pH and also the reserve pool of stored nutrients and maintained relatively stable EC level (Clark *et al.*, 1998, Gaur *et al.*, 2002) Marked variation in available N status in the mean of 2005-06 to 2007-08 (Table 19) was observed between 146.5 to 184.3 kg/ha which narrowed down between 155.05 and 168.45 kg/ha in the mean of 2008-09 to 2014-15. But the overall mean varied between 152.6 and 170.3 kg/ha being maximum in treatment T_7 where 100 per cent recommended dose of fertilizers were applied and minimum in T_5 where application of 50 per cent N through FYM along with bio-fertilizer N, rock phosphate and PSB bio-fertilizer was made.

Initial three years mean of available P varied between 12.3 and 16.5 kg/ha (Table 20) being minimum in T_5 and maximum in treatment T_7. Whereas, mean of further six years *i.e.* from 2008-09 to 2014-15 showed variation between 14.9 and 17.4 kg/ha being minimum in T_2 where 1/3 N was applied each through FYM, vermicompost and neemcake. Maximum mean available P was obsereved in T_7 where 100 per cent dose of recommended fertilizers were applied. But the overall (2005-06 to 2014-15) variation in mean available P was between 14.4 and 17.2 kg/ha. It means stabilization of available P has taken place among different treatments because of the little variation in the means of six years and overall mean with respect to available P in both means the trend remain the same *i.e.* being minimum in T_2 and maximum in T_7 along with similar variation in per cent available P. Mean available K in soil showed variation between 214.0 and 264.3 kg/ha during initial three years while

Table 20: Mean Available Phosphorus (kg/ha) of 3-years (2005-06 to 2007-08), 7-years (2008-09 to 2014-15) and Overall (2005-06 to 2014-15) among different Treatments in Organic Farming Package for Mung Bean-Wheat (*Desi*) Cropping System

Treatment		Mean Available Phosphorus		
		2005-06 to 2007-08	2008-09 to 2014-15	2005-06 to 2014-15
T_1	50 per cent Rec. NPK + 50 per cent N (FYM)	14.5	15.6	15.0
T_2	N each through FYM, V.C., Neem cake	14.3	14.9	14.4
T_3	T_2 + Intercropping (*Kharif* : Pearlmillet, *Rabi* : Mustard)	13.2	16.2	15.3
T_4	T_2 + Agro. Practices for weed and pest control	14.1	15.7	15.0
T_5	50 per cent N FYM + N bio-fertilizer + rock phosphate + PSB	12.3	16.7	16.0
T_6	T_2 + N+P bio-fertilizer	16.3	17.4	16.8
T_7	Recommended fertilizer	16.5	17.4	17.2

there after with similar trend the availability of K narrowed down between 213.5 and 234.9 kg/ha being minimum in treatment T_5 and maximum in treatment T_7 (Table 21). Overall (2005-06 to 2014-15) mean showed variation between 210.8 kg/ha in T_5 and 238.2 kg/ha in treatment T_7 (Table 21). Broadly speaking the fertility status *i.e.* OC, available N, P and K was better after the harvest of mungbean crop than after the harvest wheat (*desi*). So it can be concluded that mungbean crop can be viewed more as a soil fertility improver than as independent crops grown for their grain output. Similar results were observed by Kanwarkamla (2000). Singh *et al.* (1997) also reported that inclusion of legumes crop in its cropping system offer special advantage to farmer. Similar results have also been reported by Saroch *et al.* (2005).

Table 21: Mean Available Potash (kg/ha) of 3-years (2005-06 to 2007-08), 7-years (2008-09 to 2014-15) and Overall (2005-06 to 2014-15) among different Treatments in Organic Farming Package for Mung Bean-Wheat (*Desi*) Cropping System

Treatment		*Mean Available Potash*		
		2005-06 to 2007-08	*2008-09 to 2014-15*	*2005-06 to 2014-15*
T_1	50 per cent Rec. NPK + 50 per cent N (FYM)	227.0	223.1	220.8
T_2	N each through FYM, V.C., Neem cake	249.0	223.3	226.0
T_3	T_2 + Intercropping (*Kharif* : Pearlmillet, *Rabi* : Mustard)	255.3	223.8	227.6
T_4	T_2 + Agro. Practices for weed and pest control	242.7	220.4	223.1
T_5	50 per cent N FYM + N bio-fertilizer + rock phosphate + PSB	214.0	213.5	210.8
T_6	T_2 + N+P bio-fertilizer	253.7	227.5	230.6
T_7	Recommended fertilizer	264.3	234.9	238.2

Table 22: Regression Equations on Per cent Organic Carbon Changes under Organic Nutrient Management Package in Mungbean-Wheat (*Desi*) Cropping System

Treatments	*Regression Equations*	*R^2 Values*
T_1	$Y = -0.0002X^2 - 0.0012X + 0.4945$	0.43
T_2	$Y = -0.0032X^2 + 0.0293X + 0.446$	0.55
T_3	$Y = -0.0037X^2 + 0.0379X + 0.4195$	0.71
T_4	$Y = -0.0029X^2 + 0.0251X + 0.4688$	0.77
T_5	$Y= -0.0021X^2 + 0.0226X + 0.4079$	0.40
T_6	$Y = -0.0024X^2 + 0.0205X + 0.4652$	0.51
T_7	$Y = -0.0014X^2 + 0.0102X + 0.5045$	0.75

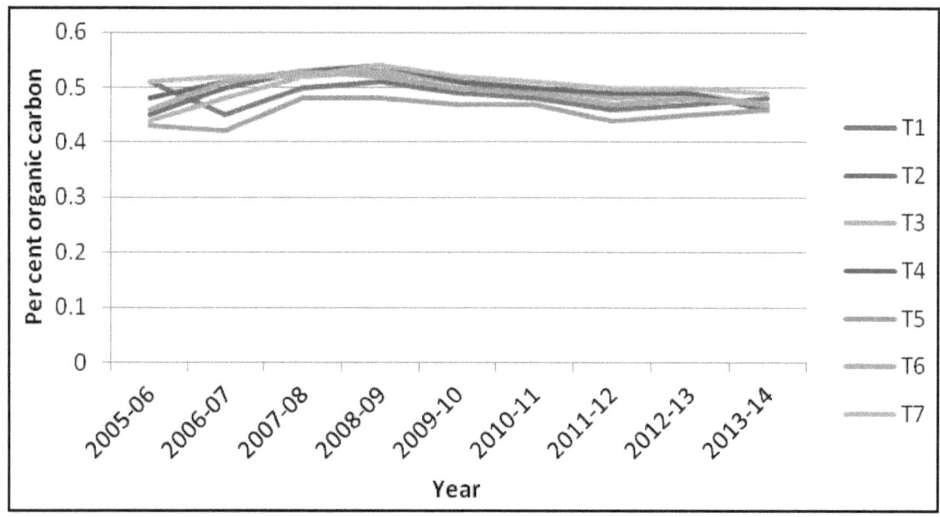

Figure 8: Regression Equations on Per cent Organic Carbon Changes under Organic Nutrient Management Packages

Table 23: Regression Equations on available Nitrogen (kg ha^{-1}) Changes under Organic Nutrient Management Package in Mungbean-Wheat (*Desi*) Cropping System

Treatments	*Regression Equations*	*R^2 Values*
T_1	$Y = 1.0106X^2 - 11.856X + 194.03$	0.73
T_2	$Y = 0.4288X^2 - 5.9912X + 176.6$	0.59
T_3	$Y= 0.8803X^2 - 10.833X + 188.53$	0.61
T_4	$Y= -0.4045X^2 + 4.2955X + 150.2$	0.39
T_5	$Y = 0.1212X^2 + 0.0945X + 145.8$	0.32
T_6	$Y = -0.4333X^2 + 4.52X + 151.97$	0.33
T_7	$Y = 1.2983X^2 - 16.756X + 214.82$	0.77

Table 24: Regression Equations on available Phosphorus (kg ha^{-1}) Changes under Organic Nutrient Management Package in Mungbean-Wheat (*Desi*) Cropping System

Treatments	*Regression Equations*	*R^2 Values*
T_1	$Y = -0.0121X^2 + 0.0945X + 14.267$	0.42
T_2	$Y = -0.0234X^2 + 0.1404X + 13.905$	0.38
T_3	$Y = -0.0948X^2 + 1.2747X + 10.895$	0.49
T_4	$Y = -0.0394X^2 + 0.4406X + 13.4$	0.38
T_5	$Y = -0.358X^2 + 4.2801X + 5.3143$	0.80
T_6	$Y = 0.0117X^2 - 0.1235X + 16.514$	0.42
T_7	$Y = -0.1281X^2 + 1.2414X + 14.829$	0.57

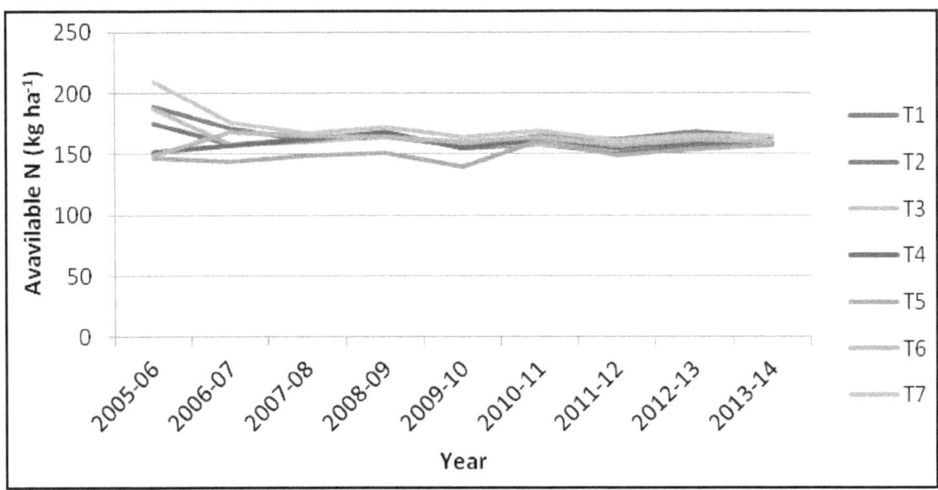

Figure 9: Regression Equations on available Nitrogen (kg ha^{-1}) Changes under Organic Nutrient Management Packages.

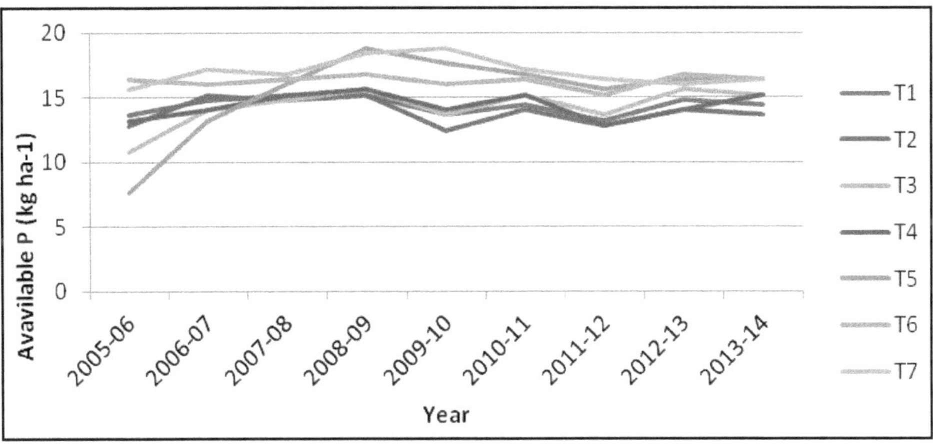

Figure 10: Regression Equations on available Phosphorus (kg ha^{-1}) Changes under Organic Nutrient Management Packages.

Major Conclusions

After completion of ten years of experimentation it is concluded that both mungbean and wheat (*Desi*) performed better with the application of recommended dose of nutrients through chemical fertilizers up to conversion period (1st-3rd crop cycles) and after conversion period (4th-10th crop cycles). System equivalent yield of 4463 kg/ha and 7236 kg/ha was recorded up to conversion period with the application of recommended nutrients through chemical fertilizers. Among organic sources of nutrients, application of 1/3 N each through FYM, vermi-compost and neem cake along with N and P bio fertilizers the system equivalent

Table 25: Regression Equations on available Potash (kg ha^{-1}) Changes under Organic Nutrient Management Package in Mungbean-Wheat (*Desi*) Cropping System

Treatments	*Regression Equations*	*R^2 Values*
T_1	$Y = 2.309X^2 - 25.091X + 270.83$	0.37
T_2	$Y = 1.9745X^2 - 26.251X + 297.31$	0.66
T_3	$Y = 2.7745X^2 - 34.998X + 318.54$	0.65
T_4	$Y = 2.3786X^2 - 28.809X + 294.42$	0.55
T_5	$Y = 2.4038X^2 - 25.96X + 261.8$	0.38
T_6	$Y = 2.6833X^2 - 32.49X + 311.17$	0.59
T_7	$Y = 3.1405X^2 - 38.368X + 333.88$	0.58

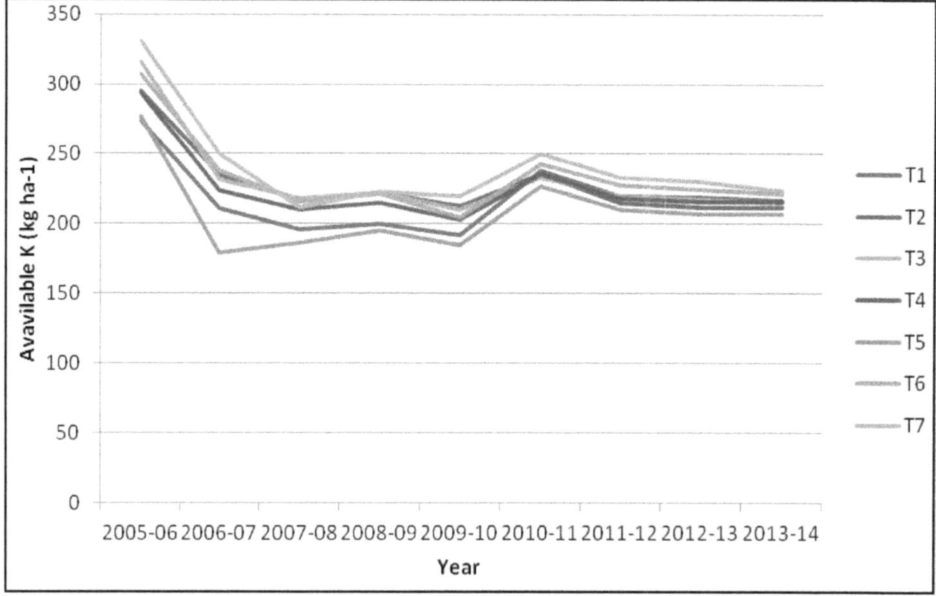

Figure 15: Regression Equations on available Potash (kg ha^{-1}) Changes under Organic Nutrient Management Packages.

yield of mungbean (929 kg/ha) and wheat (2067 kg/ha) was recorded higher. After conversion period the application of 50 per cent recommended NPK + 50 per cent N through FYM in both the crops gave the higher yield next to 100 per cent recommended fertilizers. It is interesting to note that mining of all the three nutrients *i.e.* nitrogen, phosphorus and potassium was highest in the treatment where 100 per cent recommended dose of nutrients were applied. It means that though we are getting consistently highest yield in the said treatment yet on the other hand it is at the cost of loss of nutrients from the soil. So, it can be concluded that yield enhancement with sustained soil fertility is the need of the hour for which further and continuous investigation is required.

Table 26: Changes in Soil Fertility under Organic Nutrient Management Packages in Mungbean-Wheat (*Desi*) Cropping System

Treatments	*Initial soil fertility*				*Average Soil Fertility of 10 Crop Cycles*				*Per cent I/D over initial*				*Average Change over T_7 in 10 Yrs*			
	O.C	*Av.N*	*Av.P*	*Av.K*	*O.C*	*Av.N*	*Av.P*	*Av.K*	*O.C*	*Av.N*	*Av.P*	*Av.K*	*O.C*	*Av.N*	*Av.P*	*Av.K*
	g kg^{-1}	*Kg ha^{-1}*			*gkg^{-1}*	*Kg^{-1} ha*										
T_1	5.1	189	13.6	274	4.83	166.8	14.4	218.5	5.88	13.33	-5.88	20.91	0.29	5.4	2.6	23
T_2	4.5	175	12.8	295	4.92	160.2	13.9	228.6	-4.44*	9.60	-6.25	28.14	0.20	12	3.1	12.9
T_3	4.4	187.6	10.8	316	4.91	162.2	14.3	231.4	-6.82	14.93	-40.74	32.06	0.21	10	2.7	10.1
T_4	4.8	151.2	13.2	294	5.01	158.9	14.4	225.7	4.17	-4.63	-15.15	26.63	0.11	13.3	2.6	15.8
T_5	4.3	147	7.6	277	4.56	150.1	15.4	208.1	-6.98	-6.67	-115.79	25.13	0.56	22.1	1.6	33.4
T_6	4.6	148.4	16.4	307	4.92	160.8	16.3	233.7	-2.17	-7.55	0.00	28.01	0.20	11.4	0.7	7.8
T_7	5.1	210	15.6	331	5.12	172.2	17.0	241.5	3.92	21.33	-5.13	32.33				

* Per cent increase.

Future Concern

Large-scale adoption of organic agriculture would result in food shortage as the yield reductions of organic system, as relative to conventional agriculture, is markedly high especially in intensive farming systems. However, in traditional rain-fed agriculture having 70 per cent total cultivable land, organic farming has the potential to increase the yield. Organic manure is an good alternative renewable source of nutrient supply. However, it is not possible to meet the full nutrient requirements of crops entirely from organic sources. Organic farming systems can deliver agronomic and environmental benefits both through structural changes and tactical management of farming systems. The benefits of organic farming are relevant both to developed nations (environmental protection, biodiversity enhancement, reduced energy use and CO_2 emission) and to developing countries like India (sustainable resource use, increased crop yields without over reliance on costly external inputs, environment and biodiversity protection, etc.) Organic foods are proved superior in terms of health and safely, but scientific evidence to prove their superiority in terms of taste and nutrition needs further investigation. In organic farming, pest and diseases management strategies are largely preventive rather than reactive. In general, pest and disease incidence is less severe in organic cultivation as to conventional cultivation. Organic wheat and mungbean have wide scope in our country because people are becoming more health conscious, though the decrease in yield with organic sources of nutrients was highest *i.e.* 45.37 per cent as compared to recommended fertilizers after conversion period but in return organic sources were proved helpful in maintaining the soil health. On the whole integrated use of fertilizers both organic and inorganic resulted in better soil fertility.

Summary

On the basis of ten years of experimentation (2005-06 to 2014-15) it is summarized that both mungbean as well as wheat performed better with application of recommended nutrients through chemical fertilizers as to organic sources alone, but 50 per cent NPK through chemical fertilizers along with 50 per cent N through FYM in both the crops was better than organic sources of nutrients alone. The yield level up to initial three crop cycles followed the pattern as followed during initial year of experimentation. The highest yield of mungbean was recorded 1209 kg ha^{-1} with the application of chemical fertilizers and it was followed by integration of chemical and organic sources of nutrients with the yield of 1132 kg ha^{-1}.

The wheat yield of 3007 kg ha^{-1} was recorded up to the period of conversion (1st-3rd crop cycle) with the application of chemical fertilizers and it was followed by application of 50 per cent recommended NPK + 50 per cent N through FYM with the yield of 2422 kg ha^{-1}. System equivalent yield of 4463 kg ha^{-1} was recorded with the application of 100 per cent recommended nutrients through chemical fertilizers.

The yield of mungbean followed the same trend during conversion period (4th to 10th crop cycles) as followed up to conversion period (1st-3rd crop cycle). The highest yield of mungbean 1224 and wheat 3024 kg ha^{-1} was recorded with the application of chemical fertilizers followed by the treatment where 50 per cent recommended NPK + 50 per cent N through FYM was applied with the yield of 1135 kg ha^{-1} and

2526 kg ha^{-1} in respect of mungbean and wheat. System yield of 7162 kg ha^{-1} was recorded highest after the completion of conversion period with the application of chemical fertilizers. The yield through application of organic sources (T_2 to T_6) ranged between 3384 to 5372 kg ha^{-1}, whereas integration of chemical and organic sources produced yield of 6329 kg ha^{-1}. The net income among various treatments varied between Rs. 13743 and Rs. 62612 ha^{-1}, being highest (Rs. 62612 ha^{-1}) with application of recommended chemical fertilizers. Hectolitre weight (79.78 g) and protein content (11.7 per cent) was highest with the application of recommended chemical fertilizers. Sedimentation value in organic treatments ranged between 34 and 36. On the whole integrated use of fertilizers both organic and inorganic resulted in better soil fertility. Though we get lesser yields but health conscious people prefer for the same with sustained fertility from agricultural point of view.

Acknowledgement

Authors are thankful to Indian Institute of Farming Systems Research (ICAR) Modipuram, Meerut, Director of Research and Professor and Head, Department of Agronomy, CCS Haryana Agricultural University, Hisar for providing financial support, guidance and other facilities for conduct of the experiment. The authors feel obliged to the scientists who have worked in this project since the inception of this experiment for their contribution in the conduct of this experiment.

Chapter 8

Resource Conservation Technology

Tillage and Planting Management in different Cropping Systems

The seeding technologies are complex and location and cropping system specific. Seeding technologies have direct bearing on crop (crop establishment mainly) and soil (nutrient and water availability) management. The soil tilth has to be tuned according to crop establishment requirement in a particular situation. Gupta (2003) reported that reduced tillage may result in timely planting of crop and good plant stand; moreover it saves energy and increases profit margin, but agriculture is so complex that it may not hold true in many locations. The seeding technology may even influence the weed flora over a period of time. Hobbs *et al.* (1997) has reported that reduced or zero tillage systems are often found to generate higher yields, reduce production cost and reduce erosion and other forms of land degradation, with corresponding benefits for the natural resource base in rice-wheat cropping system; however the response to reduced tillage operations may have different response in different cropping systems.

Methodology

Field experiments were conducted at the Research Farm of CCS Haryana Agricultural University, Hisar, India during 2007-08 to 2014-15. Hisar is situated at latitude 27 39′ to 30 55′ and longitude 73 8′ to 77 36′ E comprises a part of the Indo-Gangetic Plain. The net size of each plot was 8m x 5.4m. It is situated in the south-western zone of the State. It has arid tropical climate which is influenced by westerly winds in summer touching the temperature as high as 49 C, whereas in winter north-westerly winds bring down the temperature even to less than

0 C. Annual rainfall in the district varies between 300–450 mm with coefficient of variation of about 45 per cent. Annual potential evapotranspiration ranges between 1600 and 1650 mm. The soil of the experimental site was a sandy loam of medium fertility with a pH of 7.6.

Treatments

The field experiment was initiated in July 2007 with pearl millet-wheat/mustard cropping systems. The treatments consisted of four seeding technologies *viz.* T_1: zero tillage, T_2: minimum tillage (50 per cent of conventional tillage), T_3: conventional tillage, T_4: bed planting. The varieties of pearl millet, wheat and mustard taken were HHB 197, PBW 502 and RH 30, respectively. The field was prepared as per treatments. The sowing was done with a seed-fertilizer drill at recommended seed rate. In zero till treatment, the sowing was done directly with a zero till seed drill without any soil preparation after harvest of pearl millet. In bed planting, the beds were planted by a planter.

Crop Management

Pearl millet received 125 kg N ha^{-1} and 62.5 kg P_2O_5 ha^{-1} through urea and DAP. The N was applied in two doses, half at sowing and half broadcasted three weeks after crop sowing. The basal dose of phosphorus was given through DAP. Pearl millet was sown in June/July and harvested in September/October. Wheat was applied 150 kg N ha^{-1} and 60 kg P_2O_5 ha^{-1} through urea and DAP. In wheat half N was applied as basal dose and half was top dressed four weeks after crop sowing. Wheat was sown in November and harvested in April. All the crop management techniques were applied as per recommendations.

Zero Tillage.

Conventional Tillage.

Minimum Tillage.

Soil and Crop Observations

The organic carbon of top 15 cm soil layer was determined before and after the harvest of each crop. N,P,K content of grain and straw was analyzed and nutrient uptake was determined by multiplying the content with grain and straw yields.

Statistical Analysis

The data were subjected to analysis of variance for split-plot design, with methods of seeding as main plot and crops as subplots as per the procedure given by Little and Hills (1978).

Results

Grain Yield

The cropping systems had pearl millet-wheat and pearl millet-mustard under different seeding technologies. The pearl millet grain yield among different seeding technologies varied between 3317 kg ha^{-1} and 2921 kg ha^{-1} (Table 1). The pearl millet yield during all the eight years of experimentation and mean yield was highest in conventional tillage and it was 3426, 3876, 3598, 3009, 2976, 2809, 3401, 3441 and 3317 kg ha^{-1} in 2007-08, 2008-09, 2009-10, 2010-11, 2011-12, 2013-13, 2013-14, 2014-15 *Kharif* season and mean, respectively (Table 1) and it was closely followed by minimum tillage with mean yield of 3192 kg ha^{-1}. The lowest mean yield of 2921 kg ha^{-1} was recorded in the bed planting. The mean pearl millet yield was 3090 kg ha^{-1} in zero till treatment. The pearl millet yield was 3392, 3695, 3434, 2710, 2789, 2629,3194, 3170 and 3369, 3701, 3502, 2760, 2793, 2621, 3159, 3161 and 3133 kg ha^{-1} respectively in *Kharif* 2007-08, 2008-09, 2009-10, 2010-11, 2011-12, 2013-13, 2013-14, 2014-15 and mean of eight years. During *Rabi* season, the yield was highest in conventional tillage (4535 kg ha^{-1}) and lowest in bed planting (4055 kg ha^{-1}) and yield in zero tillage was 4147 kg ha^{-1} (Table 2). During *Rabi* season wheat yielded 5262 kg ha^{-1} and mustard yielded 2278 kg ha^{-1}. Similar results were obtained during all the eight years of experimentation.

Wheat Equivalent Yield

Among the tillage treatments, highest wheat equivalent yield of 5842, 7251, 7561, 6553, 7373, 7745, 13939 and 8853 kg ha^{-1} was recorded in conventional tillage during 2007-08, 2008-09, 2009-10, 2010-11, 2011-12, 2012-13, 2013-14 and 2014-15, respectively (Table3). The mean wheat equivalent yield among different seeding technologies varied between 8140 kg ha^{-1} and 7017 kg ha^{-1}; being highest with the conventional tillage and minimum in bed planting (Table 3). The wheat equivalent yield in zero tillage was 7218 kg ha^{-1} and in bed planting it was 7017 kg ha^{-1}. The wheat equivalent yield was 7532 and 6219 kg ha^{-1} in pearl millet-wheat and pearl millet-mustard cropping system, respectively.

Table 1: Effect of different Seeding Technologies on Grain Yield (kg/ha) during *Kharif* Season in different Cropping Systems

Tillage Treatment	*2007-08*	*2008-09*	*2009-10*	*2010-11*	*2011-12*	*2012-13*	*2013-14*	*2014-15*	*Mean*
Zero tillage	3322	3673	3457	2707	2819	2604	3073	3062	3090
Minimum tillage	3361	3782	3560	2726	2864	2680	3265	3300	3192
Conventional tillage	3426	3876	3598	3009	2976	2809	3401	3441	3317
Bed planting	3415	3462	3257	2497	2504	2407	2966	2860	2921
CD	NS			169.8	165	157	98	197	
Cropping System									
Pearlmillet-wheat	3392	3695	3434	2710	2789	2629	3194	3170	3127
Pearlmillet-mustard	3369	3701	3502	2760	2793	2621	3159	3161	3133
CD	NS			NS	98	101	NS	NS	

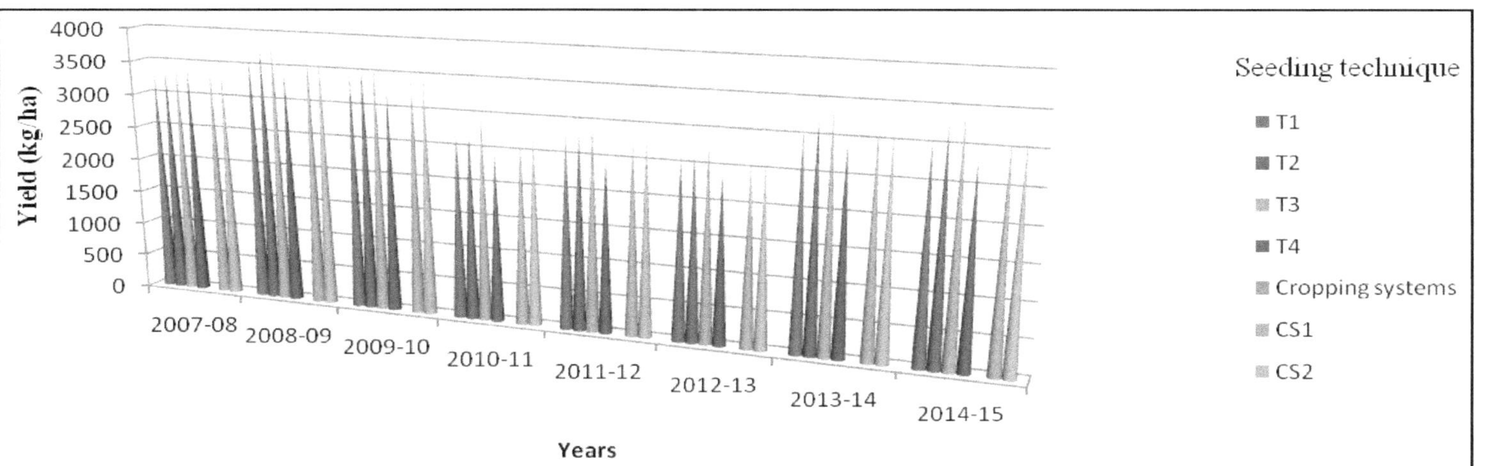

Figure 1: Effect of different Seeding Technologies on Grain Yield (kg/ha) during *Kharif* Season in different Cropping Systems.

Table 2: Effect of different Seeding Technologies on Grain Yield (kg/ha) during *Rabi* Season in different Cropping Systems

Tillage Treatment	*2007-08*	*2008-09*	*2009-10*	*2010-11*	*2011-12*	*2012-13*	*2013-14*	*2014-15*	*Mean*
Zero tillage	3562	3917	3776	3243	3571	4618	5247	5244	4147
Minimum tillage	3540	4100	3850	3415	3722	4906	5456	5288	4285
Conventional tillage	3786	4236	4017	3601	3989	5301	5572	5781	4535
Bed planting	3527	3596	3734	3108	3522	4412	5290	5250	4055
CD	NS			226.8	177	172	112	292	
Cropping system									
Pearlmillet-wheat	5404	5517	5029	5167	5327	4873	5391	5391	5262
Pearlmillet-mustard	1803	2408	2159	1517	2075	4745	1722	1796	2278
CD	NS			273.4	120	116	-	-	

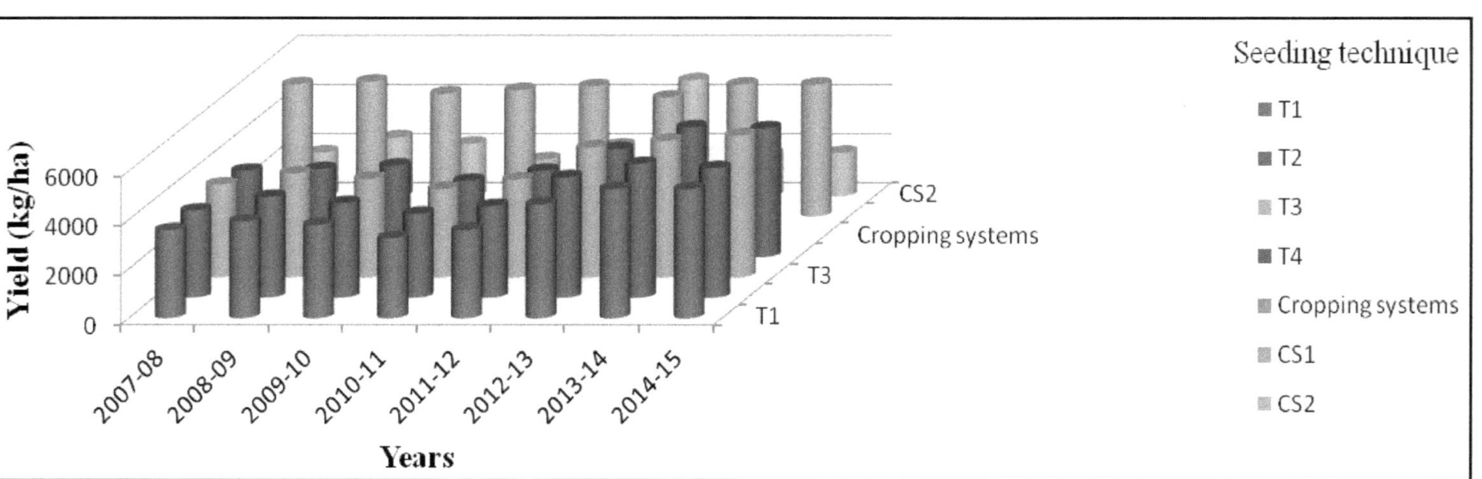

Figure 2: Effect of different Seeding Technologies on Grain Yield (kg/ha) during *Rabi* Season in different Cropping Systems.

Table 3: Effect of different Seeding Technologies on Wheat eq. Yield (WEY, kg/ha) in different Cropping Systems

Tillage Treatment	2007-08	2008-09	2009-10	2010-11	2011-12	2012-13	2013-14	2014-15	Mean
Zero tillage	5555	6773	7129	5680	6619	6882	11127	7977	7218
Minimum tillage	5557	7042	7315	5916	6908	7238	13412	8234	7703
Conventional tillage	5842	7251	7561	6553	7373	7745	13939	8853	8140
Bed planting	5576	6289	6664	5367	6314	6507	11616	7803	7017
Cropping system									
Pearlmillet-wheat	7439	8391	6056	7182	7455	7160	8348	8221	7532
Pearlmillet-mustard	5266	6898	6266	4477	6153	7205	6752	6736	6219

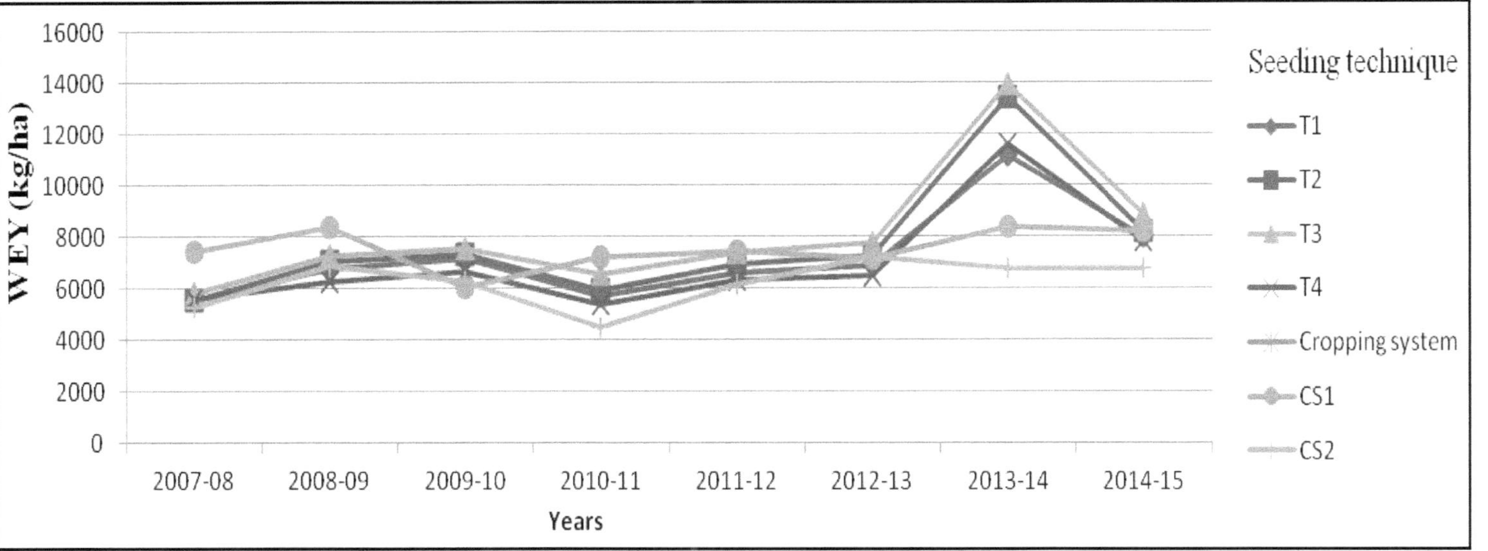

Figure 3: Effect of different Seeding Technologies on Wheat eq. Yield (WEY, kg/ha) in different Cropping Systems.

Table 4: Effect of different Seeding Technologies on Nitrogen Uptake (kg/ha) during *Kharif* Season in different Cropping Systems

Tillage Treatment	*2007-08*	*2008-09*	*2009-10*	*2010-11*	*2011-12*	*2012-13*	*2013-14*	*2014-15*	*Mean*
Zero tillage	96.1	108	94.5	79.6	68.9	67.0	73.2	73.5	82.6
Minimum tillage	98.7	115.5	103.5	84	72.3	71.6	78.3	77.4	87.7
Conventional tillage	103.6	122.6	108.2	96.5	79.5	80.7	89.3	82.9	95.4
Bed planting	100.9	108.5	92.4	79.1	65.3	68.6	72.1	71.8	82.3
Cropping system									
Pearlmillet-wheat	97	117.9	100.2	84.8	73.3	73.3	79.3	79.3	88.1
Pearlmillet-mustard	97.8	111.9	99	84.8	69.7	70.6	77.1	73.5	85.6

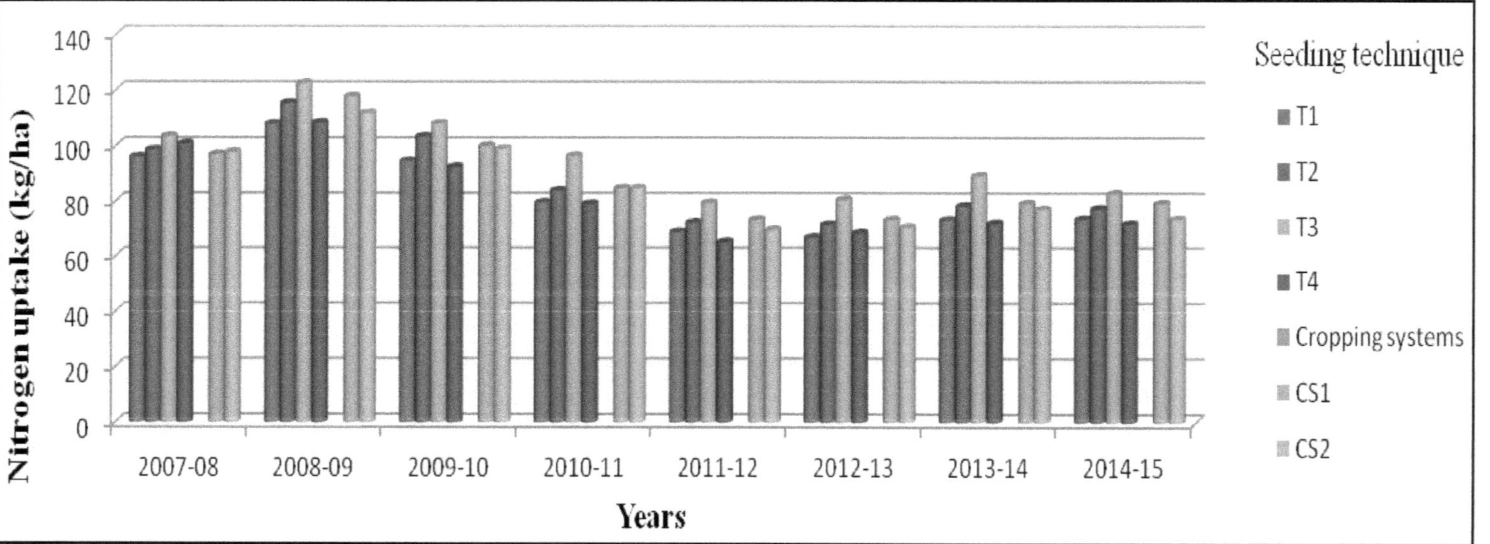

Figure 4: Effect of different Seeding Technologies on Nitrogen Uptake (kg/ha) during *Kharif* Season in different Cropping Systems.

Table 5: Effect of different Seeding Technologies on Phosphorus Uptake (kg/ha) during *Kharif* Season in different Cropping Systems

Tillage Treatment	*2007-08*	*2008-09*	*2009-10*	*2010-11*	*2011-12*	*2012-13*	*2013-14*	*2014-15*	*Mean*
Zero tillage	28.8	30.9	26.6	22.9	18.8	19.0	20.1	20.3	23.4
Minimum tillage	27.7	33.5	29	23.4	20.1	20.7	21.3	19.8	24.4
Conventional tillage	27.7	35.4	30.1	27.1	23.1	23.8	24.8	21.0	26.6
Bed planting	27.7	30.5	24.5	22.3	17.9	20.3	19.2	21.4	23.0
Cropping system									
Pearlmillet-wheat	27.8	34.4	27.7	23.8	20.7	20.6	21.1	22.1	24.8
Pearlmillet-mustard	28.1	32.4	27.4	24.1	19.3	21.3	21.6	19.2	24.2

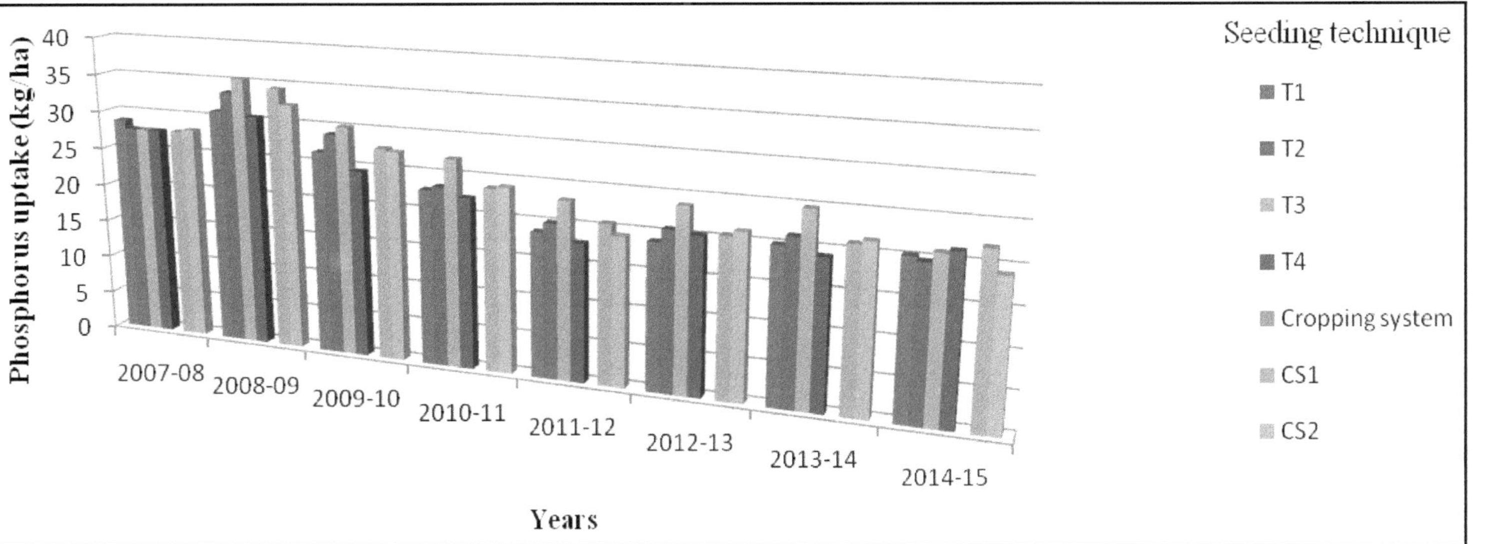

Figure 5: Effect of different Seeding Technologies on Phosphorus Uptake (kg/ha) during *Kharif* Season in different Cropping Systems.

Table 6: Effect of different Seeding Technologies on Potassium (K) Uptake (kg/ha) during *Kharif* Season in different Cropping Systems

Tillage Treatment	*2007-08*	*2008-09*	*2009-10*	*2010-11*	*2011-12*	*2012-13*	*2013-14*	*2014-15*	*Mean*
Zero tillage	232.5	249.2	224.4	196.3	183.2	179.1	180.4	174.8	202.5
Minimum tillage	234.8	255.8	238.7	199.8	191.4	188.7	182.8	167.7	207.5
Conventional tillage	240.2	272.7	240.7	221.8	209.3	205.1	206.3	175.2	221.4
Bed planting	235.4	241.4	207.3	186.3	173	180.3	175.1	189.5	198.5
Cropping System									
Pearlmillet-wheat	241.6	272.3	231	200.8	189.6	188.8	189.0	186.6	212.5
Pearlmillet-mustard	229.8	249.2	224.5	201.3	188.9	187.8	183.2	166.9	204.0

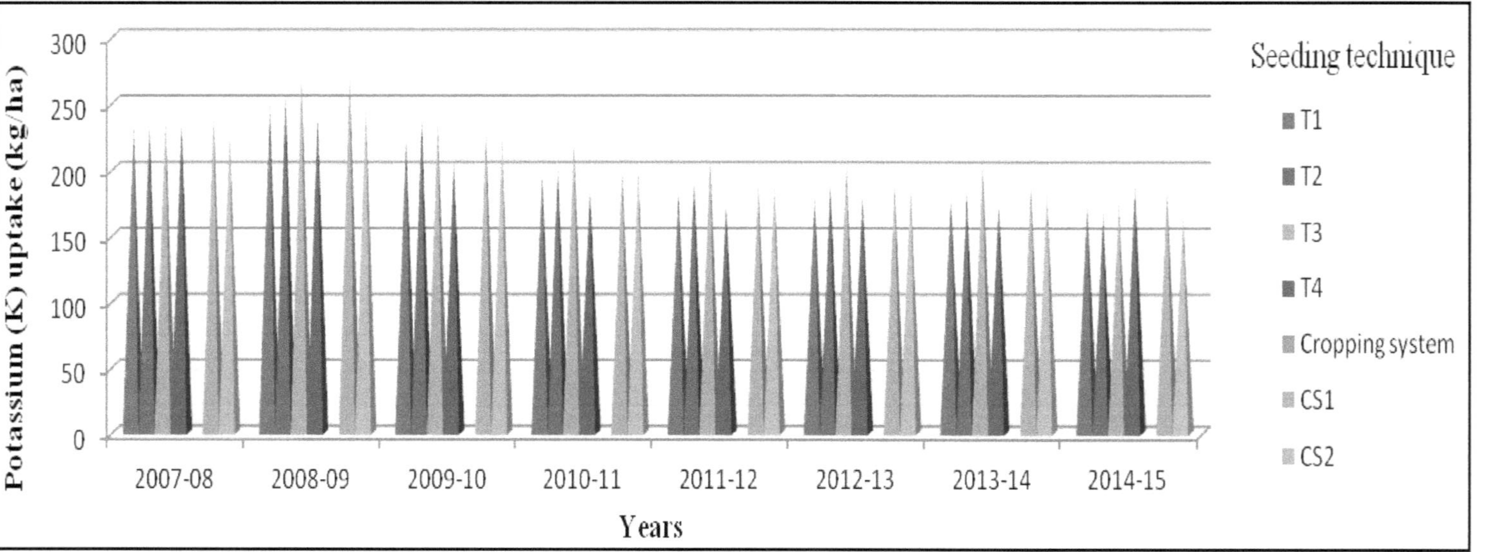

Figure 6: Effect of different Seeding Technologies on Potassium (K) Uptake (kg/ha) during *Kharif* Season in different Cropping Systems.

Table 7: Effect of different Seeding Technologies on Nitrogen Uptake (kg/ha) during *Rabi* Season in different Cropping Systems

Tillage Treatment	*2007-08*	*2008-09*	*2009-10*	*2010-11*	*2011-12*	*2012-13*	*2013-14*	*2014-15*	*Mean*
Tillage Treatment	*2007-08*	*2008-09*	*2009-10*	*2010-11*	*2011-12*	*2012-13*	*2013-14*	*2014-15*	*Mean*
Zero tillage	111.1	128.7	114.8	90.7	96.3	91.7	79.6	91.8	100.6
Minimum tillage	110.8	130.9	120.8	98.7	106.9	104.6	103.0	111.2	110.9
Conventional tillage	124.1	138.4	132.7	109.8	122.4	115.2	113.3	120.1	122.0
Bed planting	108.9	131.6	109.4	90	100.1	93.3	80.7	92.6	100.8
Cropping System									
Pearlmillet-wheat	128.3	139.5	128.5	120.9	114.7	103.2	112.3	120.0	120.9
Pearlmillet-mustard	99.1	128.3	110.3	73.7	98.3	99.2	76.0	87.8	96.6

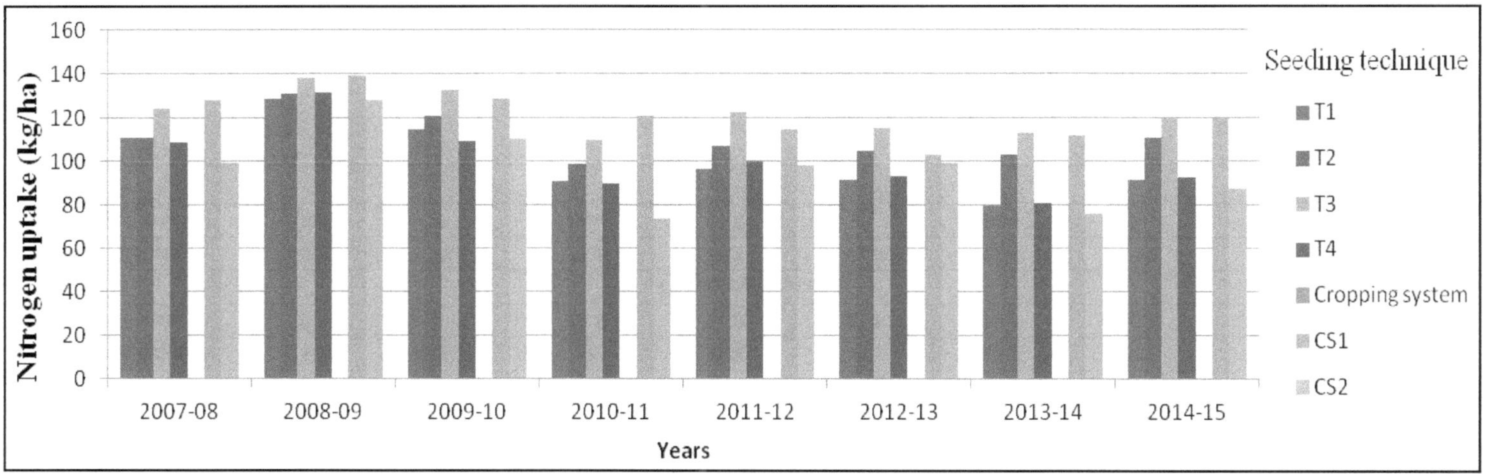

Figure 7: Effect of different Seeding Technologies on Nitrogen Uptake (kg/ha) during *Rabi* Season in different Cropping Systems.

Table 8: Effect of different Seeding Technologies on Phosphorus Uptake (kg/ha) during *Rabi* Season in different Cropping Systems

Tillage Treatment	*2007-08*	*2008-09*	*2009-10*	*2010-11*	*2011-12*	*2012-13*	*2013-14*	*2014-15*	*Mean*
Zero tillage	20.7	24.9	22.1	16.9	19.2	18.1	15.4	19.2	19.6
Minimum tillage	20.9	25.1	24.1	18.7	20.4	20.3	18.9	21.3	21.2
Conventional tillage	24.7	27.4	28	21.9	26	25.4	23.2	25.6	25.3
Bed planting	20.8	25.3	22.8	17.8	20.6	19.3	16.5	18.4	20.2
Cropping System									
Pearlmillet-wheat	19.9	23.8	21.1	19.7	19.1	17.1	19.2	20.3	20.0
Pearlmillet-mustard	23.6	26.5	27.4	17.9	24	24.4	17.7	21.9	22.9

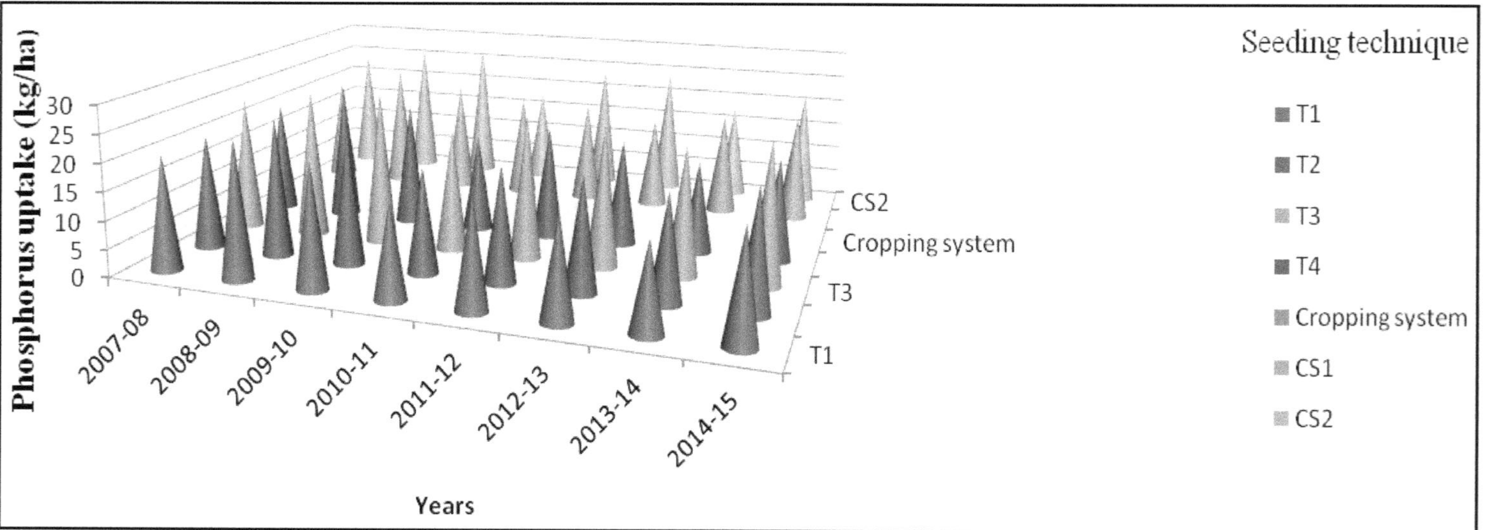

Figure 8: Effect of different Seeding Technologies on Phosphorus Uptake (kg/ha) during *Rabi* Season in different Cropping Systems.

Table 9: Effect of different Seeding Technologies on Potassium (K) Uptake (kg/ha) during *Rabi* Season in different Cropping Systems

Tillage Treatment	*2007-08*	*2008-09*	*2009-10*	*2010-11*	*2011-12*	*2012-13*	*2013-14*	*2014-15*	*Mean*
Zero tillage	123.6	129.7	131.8	104.1	106.2	102.8	80.3	111.5	111.3
Minimum tillage	126.9	133	139.2	112.1	122.1	117.8	109.1	124.9	123.1
Conventional tillage	141.5	141.5	152.2	124.7	138.1	127.9	120.8	132.0	134.8
Bed planting	124.5	133.7	126.9	102.3	110.3	104.1	87.0	115.8	113.1
Cropping System									
Pearlmillet-wheat	153.4	156.4	145.2	136.8	122.3	111.0	117.1	134.8	134.6
Pearlmillet-mustard	104.8	134.2	129.8	84.8	116.1	115.4	81.5	107.3	109.2

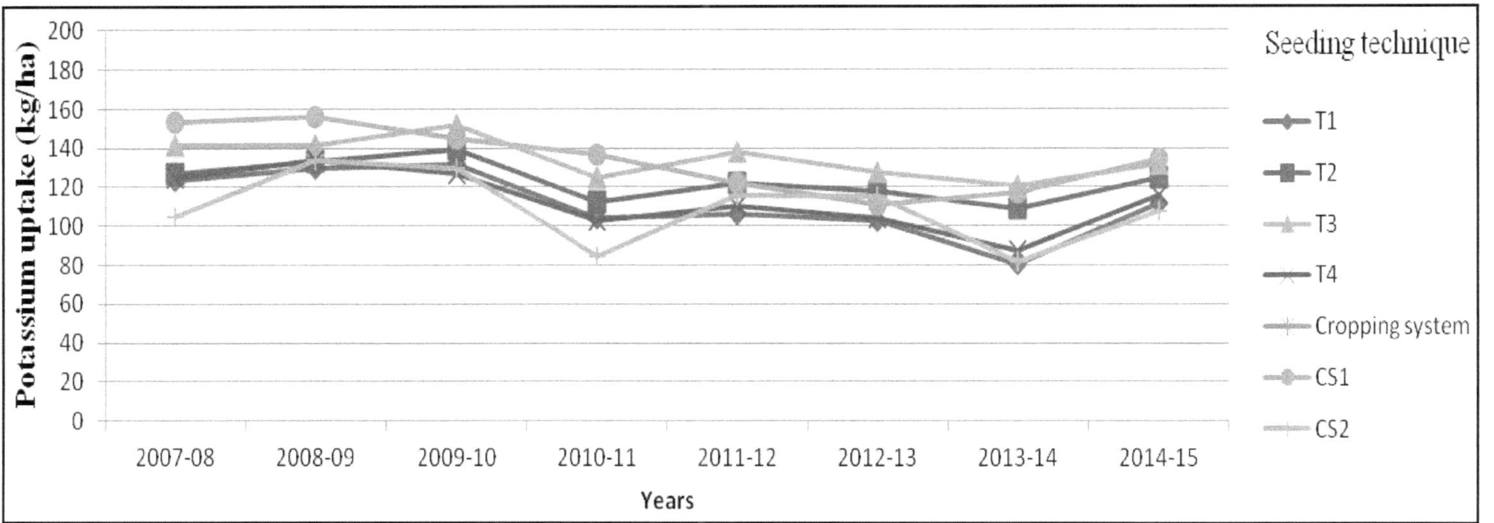

Figure 9. Effect of different Seeding Technologies on Potassium (K) Uptake (kg/ha) during *Rabi* Season in different Cropping Systems.

Table 10: Effect of different Seeding Technologies on Total {Nitrogen (N) + Phosphorus (P) + Potassium (K)} Uptake (kg/ha) during *Kharif* in different Cropping Systems

Tillage Treatment	*2007-08*	*2008-09*	*2009-10*	*2010-11*	*2011-12*	*2012-13*	*2013-14*	*2014-15*	*Mean*
Zero tillage	357.4	388.1	345.5	298.8	270.9	265.2	273.7	268.6	308.5
Minimum tillage	361.2	404.8	371.2	307.2	283.8	281.0	282.4	264.8	319.6
Conventional tillage	371.5	430.7	379	345.4	311.9	309.6	320.4	279.1	343.5
Bed planting	364	380.4	324.2	287.7	256.2	269.3	266.5	282.7	303.9
Cropping system									
Pearlmillet-wheat	346.4	424.6	358.9	309.4	283.6	282.8	289.5	288.0	322.9
Pearlmillet-mustard	355.7	393.5	350.9	310.2	277.9	279.8	282.0	259.6	313.7

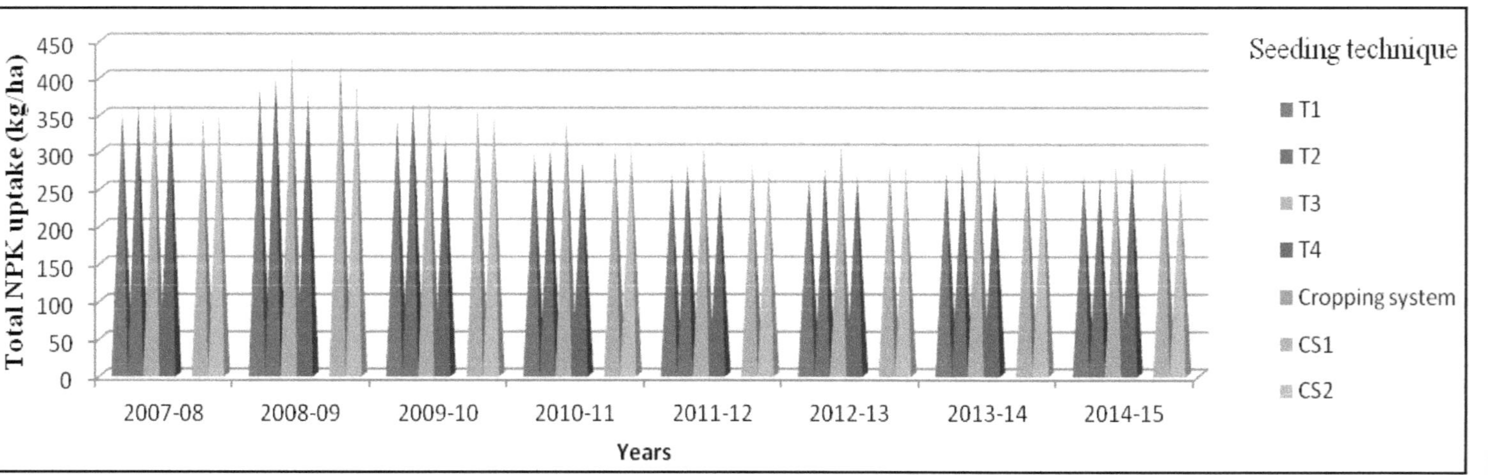

Figure 10: Effect of different Seeding Technologies on Total {Nitrogen (N) + Phosphorus (P) + Potassium (K)} Uptake (kg/ha) during *Kharif* in different Cropping Systems.

Table 11: Effect of different Seeding Technologies on Total {Nitrogen (N) + Phosphorus (P) + Potassium (K)} Uptake (kg/ha) during *Rabi* in different Cropping Systems

Tillage Treatment	*2007-08*	*2008-09*	*2009-10*	*2010-11*	*2011-12*	*2012-13*	*2013-14*	*2014-15*	*Mean*
Zero tillage	255.4	283.3	268.7	211.7	221.7	212.6	175.3	222.25	231.37
Minimum tillage	258.6	289	284.1	229.5	249.4	242.7	231.0	257.3	255.20
Conventional tillage	290.3	307.3	312.9	256.4	286.5	268.6	257.3	277.7	282.13
Bed planting	254.2	290.6	259.1	210.1	231	216.7	184.2	226.8	234.09
Cropping system									
Pearlmillet-wheat	301.6	319.7	294.8	277.4	256.1	231.3	248.6	275.1	275.58
Pearlmillet-mustard	227.5	289	267.5	176.4	238.4	239.0	175.3	217.0	228.76

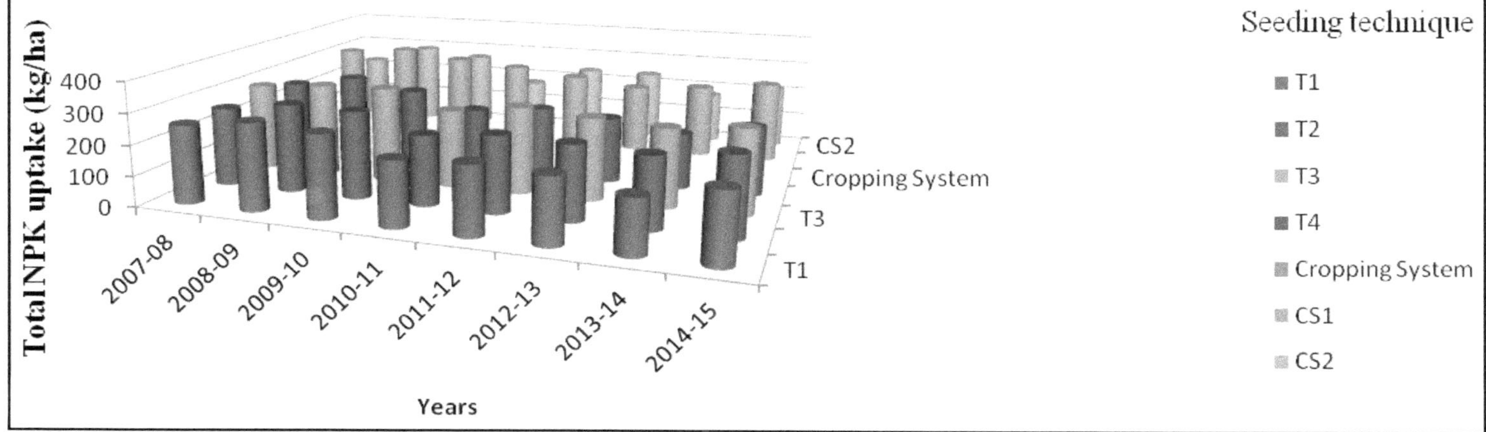

Figure 11: Effect of different Seeding Technologies on Total {Nitrogen (N) + Phosphorus (P) + Potassium (K)} Uptake (kg/ha) during *Rabi* in different Cropping Systems.

Table 12: Effect of different Seeding Technologies on Organic Carbon (Per cent) in Soil during *Kharif* Season in different Cropping Systems

Tillage Treatment	2007-08	2008-09	2009-10	2010-11	2011-12	2012-13	2013-14	2014-15	Mean
Zero tillage	0.44	0.44	0.43	0.46	0.44	0.43	0.37	0.44	0.43
Minimum tillage	0.44	0.45	0.43	0.46	0.45	0.46	0.38	0.45	0.44
Conventional tillage	0.44	0.44	0.44	0.47	0.45	0.46	0.40	0.47	0.45
Bed planting	0.44	0.44	0.44	0.45	0.43	0.43	0.36	0.44	0.43
Cropping system									
Pearlmillet-wheat	0.44	0.44	0.44	0.45	0.44	0.44	0.38	0.44	0.43
Pearlmillet-mustard	0.44	0.44	0.43	0.46	0.4	0.44	0.37	0.45	0.43

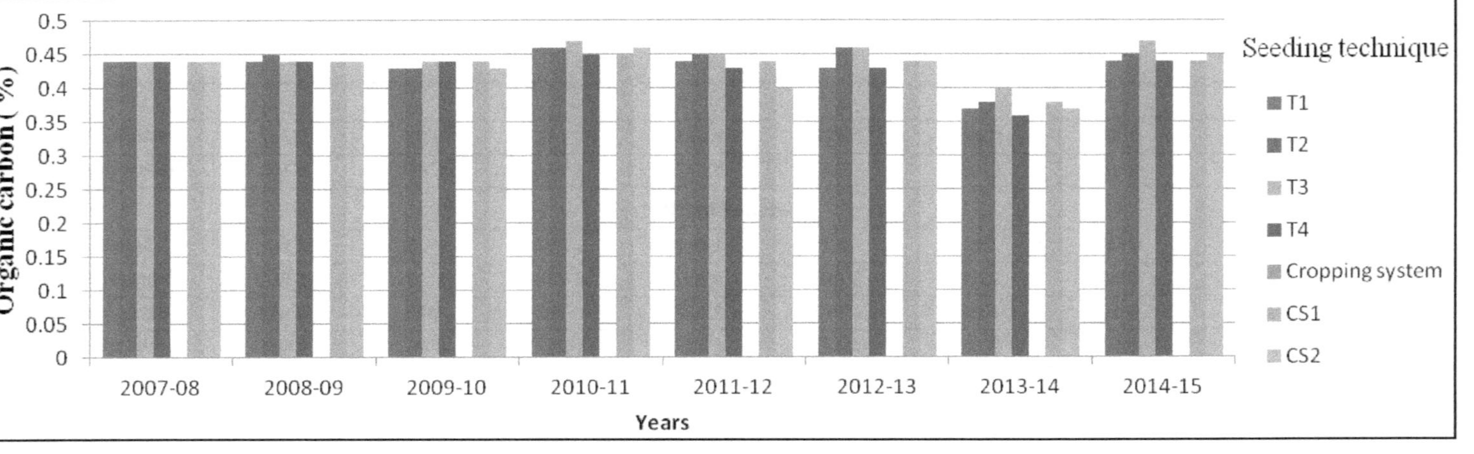

Figure 12: Effect of different Seeding Technologies on Organic Carbon (Per cent) in Soil during *Kharif* Season in different Cropping Systems.

Table 13: Effect of different Seeding Technologies on Organic Carbon (Per cent) in Soil during *Rabi* Season in different Cropping Systems

Tillage Treatment	*2007-08*	*2008-09*	*2009-10*	*2010-11*	*2011-12*	*2012-13*	*2013-14*	*2014-15*	*Mean*
Zero tillage	0.46	0.46	0.45	0.45	0.43	0.45	0.47	0.45	0.45
Minimum tillage	0.45	0.48	0.45	0.46	0.44	0.45	0.46	0.47	0.46
Conventional tillage	0.47	0.5	0.47	0.47	0.45	0.47	0.48	0.49	0.48
Bed planting	0.45	0.47	0.46	0.44	0.44	0.45	0.46	0.46	0.45
Cropping system									
Pearlmillet-wheat	0.45	0.48	0.46	0.46	0.44	0.45	0.47	0.46	0.46
Pearlmillet-mustard	0.46	0.47	0.45	0.45	0.44	0.45	0.46	0.47	0.46

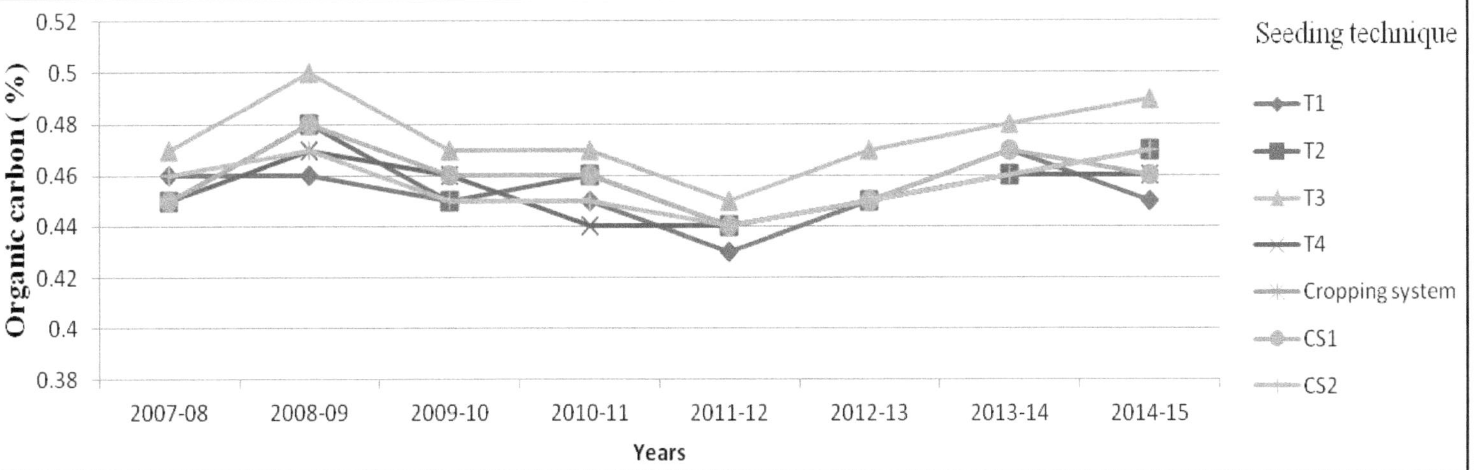

Figure 13: Effect of different Seeding Technologies on Organic Carbon (Per cent) in Soil during *Rabi* Season in different Cropping Systems.

Nutrients Uptake

During *Kharif* season the nitrogen, phosphorus and potassium uptake in zero tillage, minimum tillage, conventional tillage and bed planting was 82.6, 87.7, 95.4 and 82.3; 23.4, 24.4, 26.6 and 23.0; 202.5, 207.5, 221.4 and 198.5 kg ha^{-1}, respectively (Table 4 to 6). The pearl millet-wheat cropping system removed nitrogen, phosphorus and potassium 88.1, 24.8 and 212.5 kg ha^{-1}, while pearl millet-mustard removed 85.6, 24.2 and 204.0 kg ha^{-1}, N, P and K, respectively. During *Rabi* season the nitrogen, phosphorus and potassium uptake in zero tillage, minimum tillage, conventional tillage and bed planting was 100.6, 110.9, 122.2 and 100.8; and 19.6, 21.2, 25.3 and 20.2; and 111.3, 123.1, 134.8 and 113.1 kg ha^{-1}, respectively (Table 7 to 9). The maximum total nutrient uptake during *Kharif* was 343.5 kg and during *Rabi* 282.1 kg ha^{-1} and it was in conventional tillage during both the seasons (Table 10 and 11). During *Rabi* season, the uptake of N, P, K and total nutrients was higher in wheat as compared to mustard.

Soil Fertility Status

After the harvest of pearl millet, organic carbon (OC) status did not vary among different seeding technologies markedly (Table 12). The respective variability in OC was between 0.43 and 0.45. The organic carbon (OC) did not vary markedly among different treatments *i.e.* both seeding technologies and cropping systems during *Rabi* season also (Table 13). The variability in organic carbon among different treatments was 0.45 and 0.48 per cent.

Discussion

A eight year study indicated that conventional system of seeding gave highest grain yield and wheat equivalent yield during each year of experimentation and mean of five years. It was followed by minimum tillage and zero tillage, respectively. The lowest yield was recorded in bed planting. The higher yield in conventional tillage might have been due to better soil tilth which resulted in higher utilization of water and other inputs as a result of higher root development under such situations. The better root development ultimately caused better yield attributes and yield. Gupta *et al.* (2011) also reported higher wheat yield under conventional tillage than zero till. Pearl millet-wheat gave higher yield than pearl millet-mustard cropping system due probably to higher yielding potentiality of wheat crop.

The nutrients (N,P, K and total) uptake during all the years of experimentation was lowest in bed planting due probably to low yield level in this treatment. Gupta *et al.* (2011) also reported marginal difference in nutrients uptake between conventional and zero till treatment. The N, P, K and total nutrients uptake was similar in both cropping systems.

The physical properties of soil were not influenced markedly by both seeding technologies and cropping systems.

Conclusions

The present study conducted at CCS Haryana Agricultural University, Hisar, Haryana indicated that higher pearl millet, wheat and mustard was recorded with conventional tillage due to fine soil tilth, better root development which resulted in good aeration and better utilization of water, nutrients and other inputs. The yield level was lower in zero till and minimum tillage probably due to rough and non-uniform seed bed which resulted in lower soil nutrients, water *etc.*, utilization and ultimately lower yield in these treatments. Many workers have reported higher yield under zero tillage in many cropping systems particularly rice-wheat cropping system. Further investigation is needed under different soil types keeping important cropping systems. The bed planting gave lowest crop yields. Pearl millet-wheat gave higher crop yield than pearl millet-mustard due to higher yield potential of wheat crop. In this experiment plot size should be bigger so that mechanized manipulations can be done easily and will be more effective. Residue management is another important issue which needs to be addressed in the experiment. At least one third residue must be kept in the field. Crops should be harvested a bit higher than the ground level. Weeds should be controlled by applying recommended weedicides like glyphosate. Further research is needed on these aspects also.

Acknowledgement

Authors are thankful to Indian Institute of Farming Systems Research (ICAR) Modipuram, Meerut, Director of Research and Professor and Head, Department of Agronomy, CCS Haryana Agricultural University, Hisar for providing financial support, guidance and other facilities for conduct of the experiment. The authors feel obliged to the scientists who have worked in this project since the inception of this experiment for their scientific contribution.

Chapter 9

On Farm Trials

Significance of the Programme

Under the aegis of AICRP on FSR, three experiments namely "On farm response to application of major plant nutrients in cotton-wheat cropping system", "Diversification of existing farming systems under marginal household conditions" and "On farm evaluation of farming system modules for improving the profitability of small and marginal farmers" were initiated during the year 2011-12 and is continued till date under on farm research in Sirsa district.

The first experiment of on farm research assumes significance in the light of response of nutrients in pre-dominant cropping system of north-west zone of the state *i.e.* cotton-wheat specially, it signifies the role of nutrients in the increasing the food production. The most commonly observed effect of intensive cereal based cropping system is a decline in the partial factor productivity of nitrogen fertilizer (Hobbs and Morris, 1996). Declining soil nitrogen supply results in declining factor productivity of chemical nitrogen, since soil nitrogen is a natural substitute for chemical nitrogen. The magnitude of 30 per cent yield decline over a 20 year period due to decaling soil nitrogen supply is estimated by Cassman and Pingali (1993). In addition to nitrogen, phosphorus and potassium are the two other macro-nutrients demanded by many cropping system prevailing in Haryana. Phosphorus and potassium deficiencies are becoming widespread in areas not previously considered to be deficient. These deficiencies are directly related to the increase in cropping intensity and the predominance of year-round irrigated production systems.

Second experiment on diversification of existing farming systems is also very important as it will enhance the productivity and profitability of marginal farmers through IFS approach. Also it will estimate the component of capacity building in diversification of crop + livestock system, In this experiment emphasis was given on crop diversification, livestock diversification, product diversification and capacity building *i.e.* training of farmers. These four modules have given encouraging results and have improved the net income of the adopted farm family.

The third experiment addresses the critical constraints in small farm system for overall improvement. It will also ensure the profitability of households and livelihood security. The goal of the above experiment was achieved through removing the constraints of crop component *i.e.* using of improved varieties and applying recommended fertilizers. Critical components of animal component are being taken up through balanced feeding. Value addition of farm produce for more profit is another option. Besides, nutritional kitchen gardening was also introduced in the backyard of the house.

On Farm Response to Application of Major Plant Nutrients in Cotton-Wheat Cropping System

Cotton-wheat is an important cropping system of south west zone of Haryana. Farmers usually do not apply balanced fertilization to both the crops. This not only produce potential yield of this cropping system but also have adverse impact of crop-livestock continuum. This experiment was specially planed to estimate the impact of nutrient deficiencies in the soil and to harvest optimum yield by applying balanced fertilization in this important cropping system. Keeping in view the above aspects in mind the experiment was conducted with following objectives:

- ✰ To assess the response of major crops to nutrients in predominant cropping systems in different agro-ecosystems.
- ✰ To estimate the impact of nutrient deficiency on crop-livestock-human continuum.

Methodology

The experiment was conducted at farmer's field in Rania and Ellenabad blocks of Sirsa district in six villages at 24 locations. The experiment was started during *Kharif* 2011. The details of the treatments are given as below:

Treatment	
T_1	Control (no fertilizer)
T_2	N (Recommended dose of N for crop component in the crop sequence)
T_3	NP (Recommended dose of N and P for crop component in the crop sequence)
T_4	NK (Recommended dose of N and K for crop component in the crop sequence)
T_5	NPK (Recommended dose of N,P and K for crop component in the crop sequence)
T_6	NPK (Recommended dose of N,P and K for crop component in the crop sequence) + Micro nutrient ($ZnSO_4$)
T_7	Farmer's practice

Gross plot size :	500 m^2
Net plot size :	100 m^2
Design :	RBD

Recommended dose of cotton kg ha^{-1} (NPK) = 175:60:60

Recommended dose of wheat kg ha^{-1} (NPK) = 150:60:60

Experimental Results and Research Achievements

Yield and Economics

The mean data in Table 1 indicate that in cotton crop, the seed cotton yield was the lowest 1011, 1115, 1163, 758, 460 and 901 kg ha^{-1} in control plot (T_1) where no fertilizer was applied during 2011-12, 2012-13, 2013-14, 2014-15, 2015-16 and mean, respectively. Seed cotton yield was found to be increased with the application of 175 kg N ha^{-1} or its combinations with P, K or both. The treatment T_6 gave maximum yield of 2875, 2759, 2381, 2263, 1897 and 2435 kg ha^{-1} during 2011-12, 2012-13, 2013-14, 2014-15, 2015-16 and mean, respectively, which shows the significant response of zinc sulphate application. During *Rabi* season, the wheat yield followed cotton lint yield pattern in all the treatments. The maximum wheat yield of 5649, 5109, 5481, 4099, 4963 and 5060 kg ha^{-1} was recorded in T_6 treatment during all the years of study and mean. It was followed by T_5, T_7, T_3, T_4, T_2 and T_1 in descending order of magnitude. The lowest wheat yield of 1691, 1394, 1398, 867, 1095 and 1289 kg ha^{-1} was recorded in control plot (T_1) where no fertilizer was applied during 2011-12, 2012-13, 2013-14, 2014-15, 2015-16 and mean, respectively.

The gross return and return over variable cost/net return of the system based on five year experimentation were the highest in T_6 (Rs. 193164 ha^{-1} $annum^{-1}$) and (Rs. 111934 ha^{-1} $annum^{-1}$), respectively (Table 2). It was followed by T_5 (Rs. 182124 ha^{-1} $annum^{-1}$) and (Rs. 103951 ha^{-1} $annum^{-1}$), respectively. The lowest gross return and return over variable cost/net returns were registered in control plot (Rs. 63440 ha^{-1} $annum^{-1}$) and (Rs. -839 ha^{-1} $annum^{-1}$), respectively.

The benefit cost ratio was the highest (2.41) with T_6 followed by T_5 (2.33), T_7 (2.22), T_3 (2.22), T_4 (2.15), T_2 (1.90) and the lowest with T_1 (1.00).

Nutrient Uptake

There was differential response in nutrient uptake due to different treatments (Table 3 and 4). The range of total nutrient uptake in cotton crop was 89.7 to 276.8 kg ha^{-1} and in wheat crop 63.6 to 287.7 kg ha^{-1} with the different treatments. Nutrient uptake increased with the additional application of nutrients being highest with T_6 where recommended rates of NPK with zinc sulphate were applied. The higher uptake of nutrients in NPK with zinc sulphate over other treatments was attributed to higher grain and straw yields of cotton and wheat crops. In cotton, the maximum total uptake of nutrient (N+P+K) was 374.08, 320.35, 254.07, 243.39, 191.93 and 276.8 kg ha^{-1} was analyzed in T_6 treatment during 2011-12, 2012-13, 2013-14, 2014-15, 2015-16 and mean, respectively, whereas the minimum 111.05, 113.0, 116.92, 68.18, 39.24

Visit of Scientists of OFR Sirsa in the Experimental Field in Village Naharana.

Table 1: Effect of Nutrient Management Treatments on Yield (kg/ha) of Cotton and Wheat in Cotton-Wheat System

Treatment	*2011-12*		*2012-13*		*2013-14*		*2014-15*		*2015-16*		*Mean*	
	Cotton	*Wheat*	*Cotton*	*Wheat*	*Cotton*	*Wheat*	*Cotton*	*Wheat*	*Cotton*	*Wheat*	*Cotton*	*Wheat*
T1	1011	1691	1115	1394	1163	1398	758	867	460	1095	901	1289
T2	2063	4722	1901	3768	1616	3767	1401	2168	1177	2928	1632	3471
T3	2386	5176	2390	4241	2065	4917	1949	3576	1499	4352	2058	4452
T4	2270	4828	2200	4096	1778	4210	1655	3322	1326	4179	1846	4127
T5	2729	5448	2641	4922	2202	5249	2110	3896	1686	4689	2274	4841
T6	2875	5649	2759	5109	2381	5481	2263	4099	1897	4963	2435	5060
T7	2482	5260	2388	4663	2059	4903	1944	3623	1590	4340	2093	4558
CD	134	221	114	193	69	161	82.5	158	107.5	157.6		

Details of treatments is as follows:

T_1: Control; T_2: Nitrogen (N); T_3: Nitrogen (N) + Phosphorus (P); T_4: Nitrogen (N) + Potassium (K); T_5: Nitrogen (N) + Phosphorus (P) + Potassium (K); T_6: Nitrogen (N) + Phosphorus (P) + Potassium (K) + Zinc (Zn); T_7: Farmer's Practice.

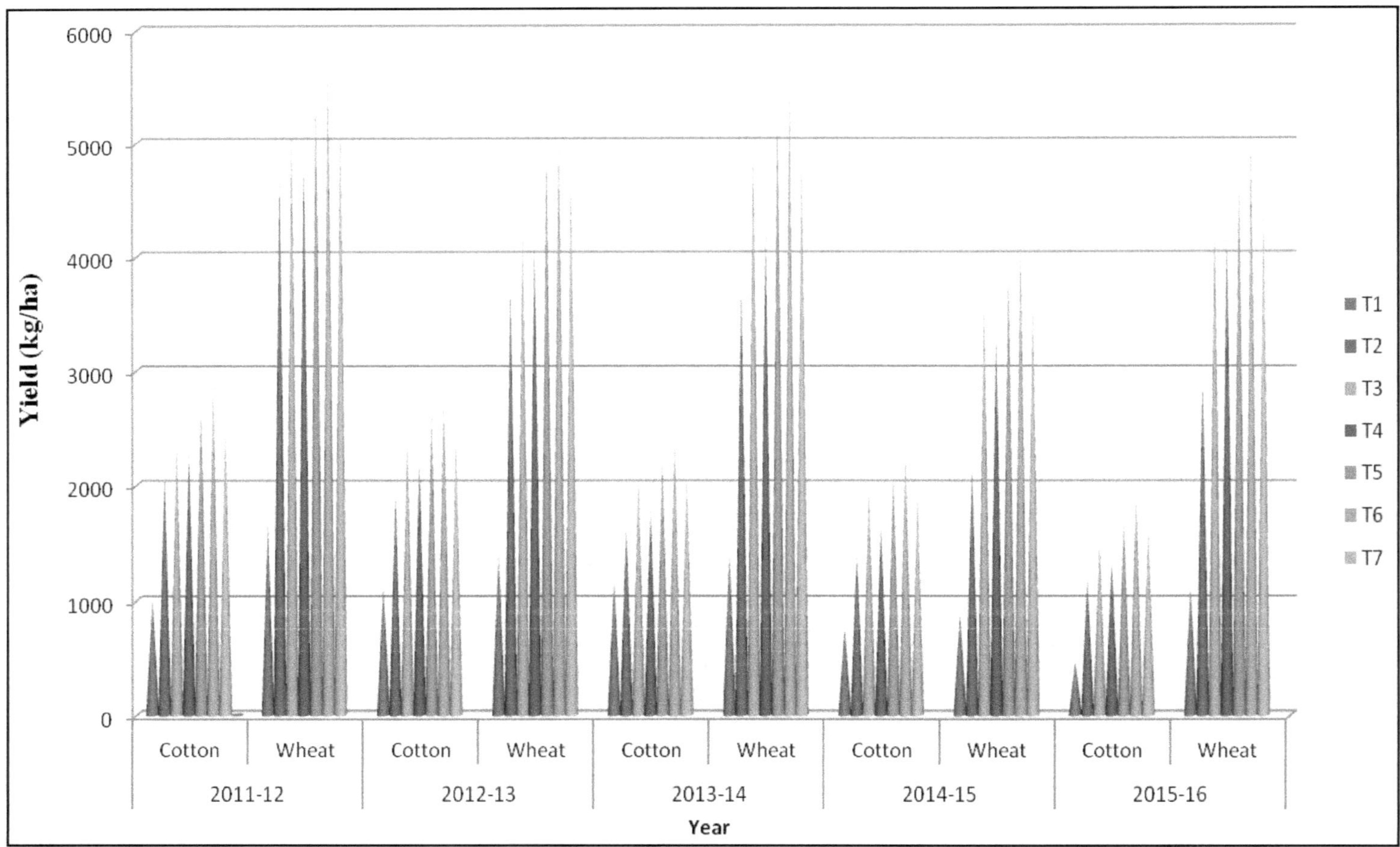

Figure 1: Effect of Nutrient Management Treatments on Yield (kg/ha) of Cotton and Wheat in Cotton-Wheat System.

Table 2: Effect of Nutrient Management Treatments on Economics in Cotton-Wheat System

Treat-ment	2011-12			2012-13			2013-14			2014-15			2015-16			Mean		
	Gross Returns	Return over Variable Cost/Net Returns	B:C Ratio	Gross Returns	Return over Variable Cost/Net Returns	B:C Ratio	Gross Returns	Return over Variable Cost/Net Returns	B:C Ratio	Gross Returns	Return over Variable Cost/Net Returns	B:C Ratio	Gross Returns	Return over Variable Cost/Net Returns	B:C Ratio	Gross Returns	Return over Variable Cost/Net Returns	B:C Ratio
T1	68943	5070	1.07	69828	761	1.12	90710	28637	1.45	47446	-15339	0.75	40275	-23325	0.63	63440	-839	1.00
T2	160150	92627	2.37	141602	64974	2.13	148147	80022	2.17	97711	28901	1.42	105576	36116	1.51	130637	60528	1.92
T3	180799	108700	2.51	170985	97690	2.33	188035	113363	2.49	141179	65844	1.87	145170	69235	1.90	165234	90966	2.22
T4	170774	100751	2.44	159332	78244	2.27	168525	92526	2.37	130151	58181	1.81	134580	61742	1.84	152672	78289	2.15
T5	202098	127499	2.71	191212	113410	2.46	196816	118468	2.51	160832	81524	2.02	159662	78854	1.97	182124	103951	2.33
T6	209691	133792	2.77	199280	118585	2.47	216707	134695	2.64	166548	83538	2.10	173594	89060	2.05	193164	111934	2.41
T7	185114	112400	2.55	176609	100764	2.33	191259	113523	2.46	146481	68931	1.89	148988	69684	1.87	169690	93060	2.22

Table 3: Total Nutrient Uptake (kgha^{-1}) as Influenced by different Treatments in Cotton

Treatments	2011-12	2012-13	2013-14	2014-15	2015-16	Mean
T1	111.05	113.00	116.92	68.18	39.24	89.7
T2	240.24	200.91	159.75	132.73	106.97	168.1
T3	294.19	269.86	207.90	194.97	135.14	220.4
T4	272.91	233.04	195.25	161.55	122.59	197.1
T5	349.73	296.16	230.41	216.50	155.28	249.6
T6	374.08	320.35	254.07	243.39	191.93	276.8
T7	305.90	265.16	213.88	192.99	151.58	225.9
CD	16.90	12.00	6.98	8.94	14.43	

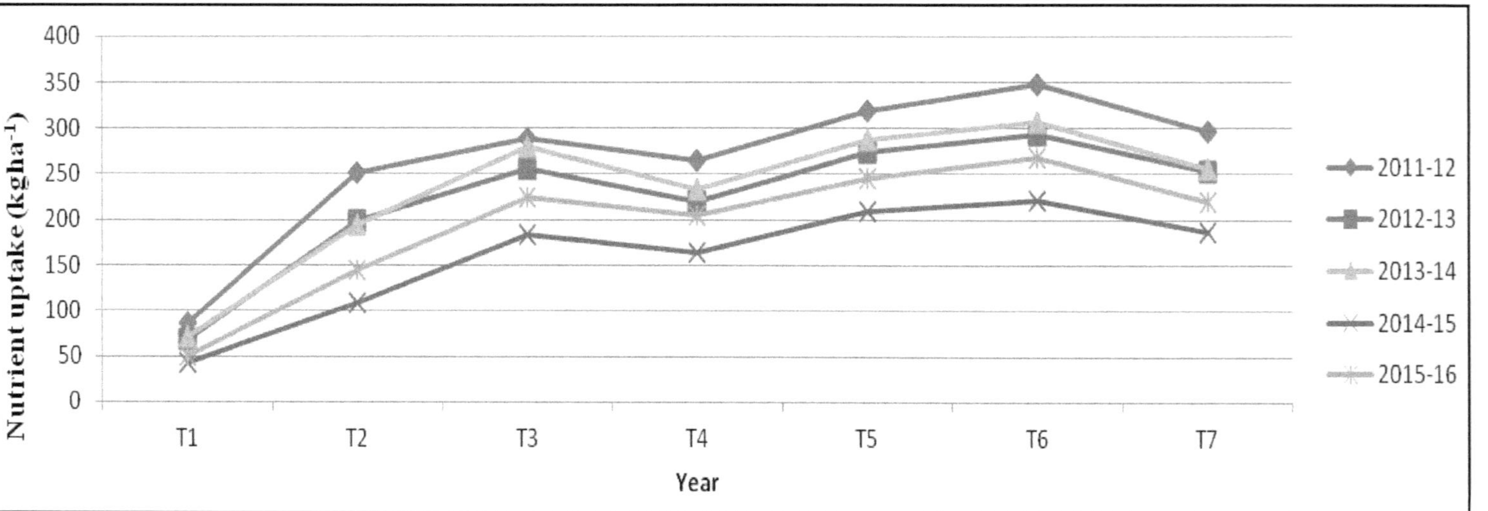

Figure 2: Total Nutrient Uptake (kgha^{-1}) as Influenced by different Treatments in Cotton.

Table 4: Total Nutrient Uptake ($kgha^{-1}$) as Influenced by different Treatments in Wheat

Treatments	*2011-12*	*2012-13*	*2013-14*	*2014-15*	*2015-16*	*Mean*
T1	85.48	68.49	70.56	42.70	50.60	63.6
T2	252.06	198.46	193.75	108.10	145.19	179.5
T3	289.19	256.42	279.35	183.72	223.70	246.5
T4	265.45	220.31	233.82	164.93	204.67	217.8
T5	319.12	274.44	287.2	209.69	246.06	267.3
T6	347.49	293.89	306.93	221.66	268.52	287.7
T7	297.14	253.30	256.09	187.02	220.00	242.7
CD	13.00	10.06	13.88	10.13	9.40	

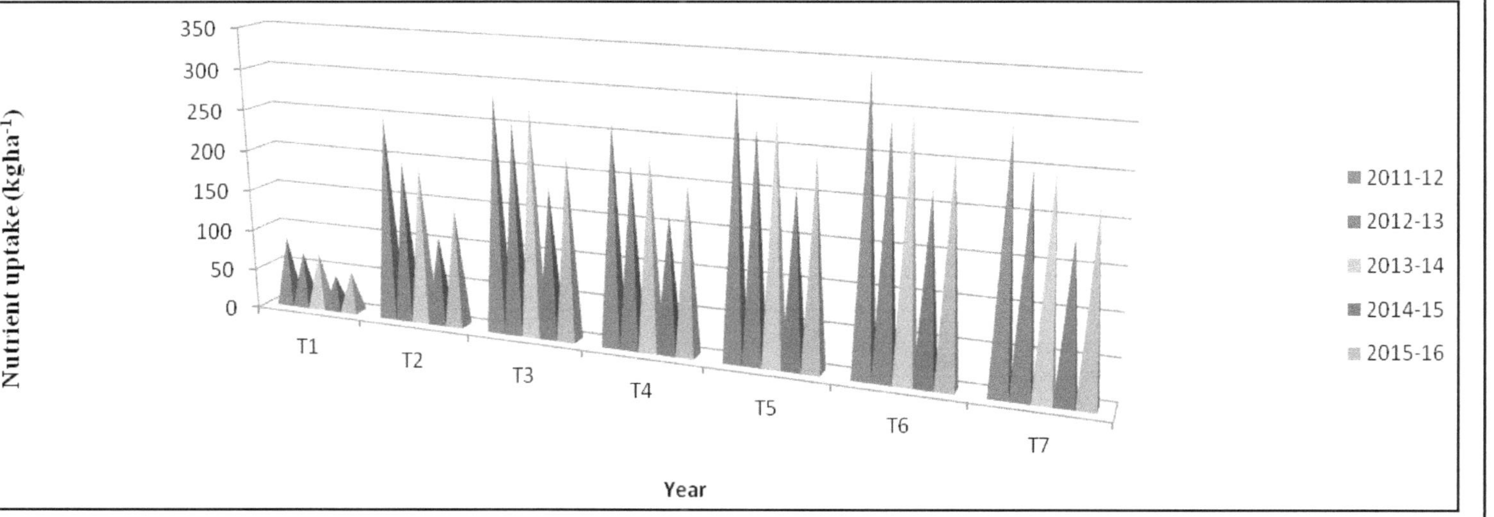

Figure 3: Total Nutrient Uptake ($kgha^{-1}$) as influenced by different Treatments in Wheat.

Table 5: Nutrients Response (kg/ha) in Cotton and Wheat as Influenced by different Treatments in Cotton-Wheat Cropping System

Treatments	*Cotton*			*Wheat*		
Years	*N*	*P*	*K*	*N*	*P*	*K*
2011-12	6.01	5.38	3.45	20.20	7.57	1.77
2012-13	4.49	8.15	4.98	15.83	7.88	5.47
2013-14	2.58	7.48	2.70	15.79	19.16	7.38
2014-15	3.67	9.13	4.23	8.67	23.46	19.23
2015-16	4.10	5.48	2.48	12.22	23.73	20.85
Mean	4.17	7.124	3.568	14.542	16.36	10.94

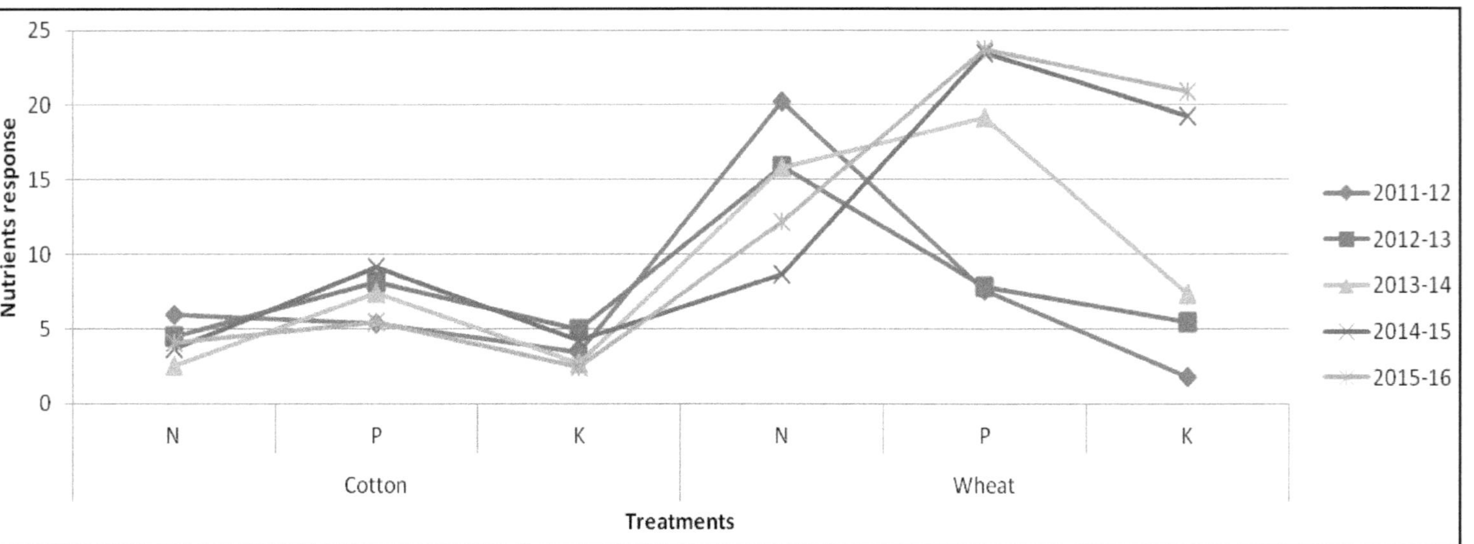

Figure 4: Nutrients Response in Cotton and Wheat as Influenced by different Treatments in Cotton-Wheat System.

and 89.7 was recorded in treatment T_1 during 2011-12, 2012-13, 2013-14, 2014-15, 2015-16 and mean, respectively.

In wheat crop the total uptake of nutrient (N+P+K) was 347.49, 293.89, 306.93, 221.66, 268.52 and 287.7 kg ha^{-1} was analyzed in T_6 treatment during 2011-12, 2012-13, 2013-14, 2014-15, 2015-16 and mean, respectively.

Nutrient Response

The data in Table 5 reveal that mean response of nitrogen, phosphorus and potash was higher in wheat crop than in cotton crop. Application of nitrogen increased more yield as compared to phosphorus and potassium in wheat crop and application of phosphorus was more effective in increasing the yield as compared to potassium in cotton-wheat system. The response in yield by applying 1.0 kg N in cotton and wheat was 4.17 and 14.54 kg ha^{-1}, respectively. Similarly, the values for phosphorus were 7.12 and 16.36 kg ha^{-1} and for potassium the values were 3.56 and 10.94 kg ha^{-1}, respectively.

Diversification of Existing Farming Systems under Marginal Household Conditions

The increased cropping intensity will be a key strategy for future gains in crop production. Short duration pulses, oilseeds and other high value crops will find their definite niche as sequential or intercrops, rather than replacing the major cereal crops having higher yield stability. Hence, an increased cropping intensity will contribute substantially to additional demands of food and cash crops. Development of new crop varieties with more efficient photosynthetic apparatus and shorter duration would be of immense help in increasing cropping intensity. Similarly, bio-intensive diversified complementary cropping systems would enable small and marginal farmers to utilize limited land and water resources in more efficient manner. The diversified cropping systems need to be considered in farming system perspective. Keeping in view the above aspects in mind the experiment was conducted with following objectives:

- ☆ To enhance the productivity and profitability of marginal farmers households through IFS approach.
- ☆ To improve the livelihood and nutritional security through diversification approach.
- ☆ To estimate the impact of capacity building in diversification of crop + livestock system.

Methodology

The experiment was conducted at farmer's field in Rania and Ellenabad blocks of Sirsa district in six villages at 24 locations. The experiment was started during *Kharif* 2013. The details of the treatments are given as below:

Details of the Treatments

M_0: Bench mark survey of household

M_1: Crop diversification

M_2: Livestock diversification

M_3: Product diversification

M_4: Capacity building (Training of farmers)

M_1 (Module-1): Crop Diversification

The cotton- wheat cropping system was diversified to guar-wheat cropping system in 0.1 ha area at farmer's field. The yield of cotton, wheat, guar and wheat was 1970, 4380, 1355 and 4525 kg/ha during the year 2013-14, respectively. During the year 2014-15 the yield of cotton, wheat, guar and wheat was 2169, 3969, 1147 and 4184 kg/ha respectively. The average net income from guar crop (42200 and 28920) was less than cotton crop (49020 and 38162) during 2013-14 and 2014-15, respectively. The average net income from cotton- wheat cropping system was higher than guar-wheat cropping system. However, the B: C ratio in guar (3.02 and 2.28) was higher than cotton (1.95 and 1.74) during 2013-14 and 2014-15, respectively as shown in Table 1.

M_2 (Module -2): Livestock Diversification

The goat rearing was introduced as livestock diversification during 2013-14. Less number of farmers agreed to adopt this enterprise due to social reasons. The goat rearing is in progress at some of the household. The farmer got average profit of Rs 5049 from goat rearing in one year.

M_3 (Module-3): Product diversification

Nutritional kitchen gardening was started in small area with the idea to save farmer's expenditure on purchasing vegetables for home consumption. The average saving of farmer by nutritional kitchen gardening was Rs 3095 per farmer in one year of vegetable cultivation. The average profit in ghee making was Rs 9065 per farmer per year.

M_4 (Module-4): Capacity Building

Five trainings were imparted to the marginal farmers of selected villages on nursery raising, proper spray techniques, crop production and kitchen gardening.

On - Farm Evaluation of Farming System Modules for Improving the Profitability of Small and Marginal Farmers

The prevailing farming situation in India calls for an integrated effort to address the emerging issues/problems. The integrated farming systems approach is considered to be the most powerful tool for enhancing profitability of farming systems, especially for small and marginal farm-holders to make them bountiful. In fact, our past experience has clearly evinced that the income from cropping alone is hardly sufficient to sustain the farmers' needs. With enhanced consumerism in

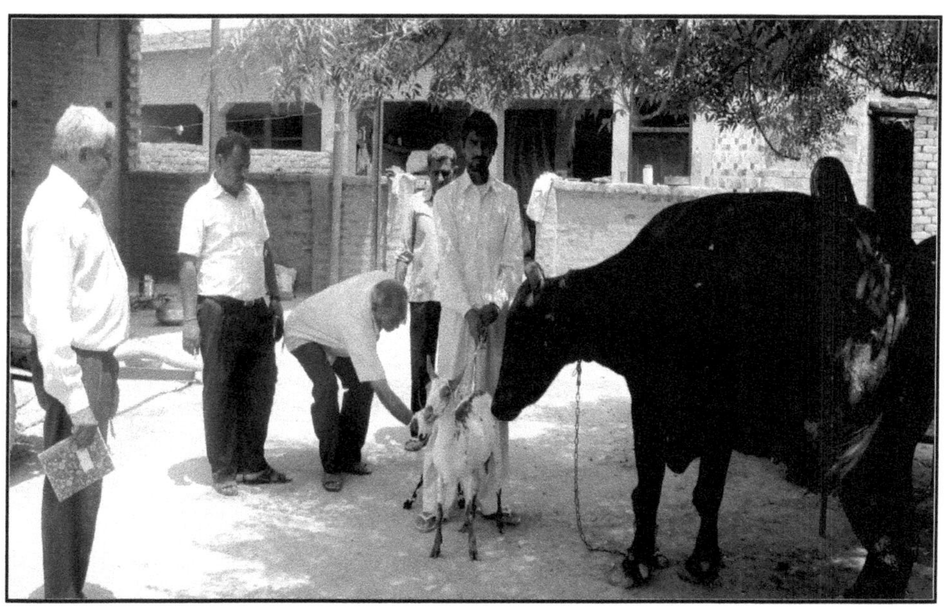

Livestock Diversification in OFR.

Table 1: Yield and Economics of Existing Crop with Diversified Crop

Crop	*2013-14*				*2014-15*			
	Cotton	*Wheat*	*Guar*	*Wheat*	*Cotton*	*Wheat*	*Guar*	*Wheat*
Seed cotton/Grain Yield (kg ha^{-1})	1970	4380	1355	4525	2169	3969	1147	4184
Straw Yield (kg ha^{-1})	4531	4420	1480	4670	5136	4072	1175	4293
Gross return (Rs. ha^{-1})	100310	72370	63000	75025	89702	71802	51470	75693
Cost of cultivation (Rs. ha^{-1})	51290	32735	20800	32735	51540	33880	22550	33880
Net Return (Rs. ha^{-1})	49020	39635	42200	42290	38162	37922	28920	41813
B:C ratio	1.95	2.21	3.02	2.29	1.74	2.12	2.28	2.23

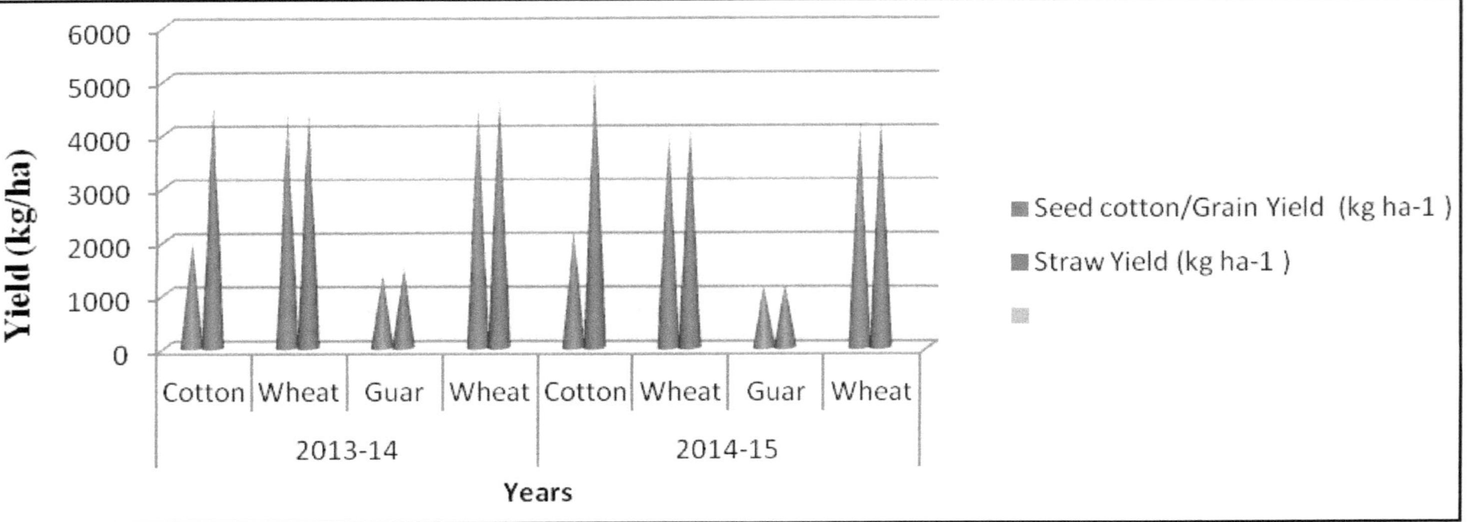

Figure 1: Yield of Existing Crop with Diversified Crop.

rural areas, farmers' requirements for cash have also increased to improve their standard of living. This is especially true in case of small and marginal farmers. Therefore, farmers' income and food requirements would have to be augmented and supplemented by adoption of efficient secondary/tertiary enterprises like animal husbandry, horticulture (vegetables/fruits/flowers/medicinal and aromatic plants), apiary, mushroom cultivation, fisheries etc. However, these integrated farming systems will be required to be tailor-made and designed in such a manner that they lead to substantial improvement in energy efficiencies at the farm and help in maximum exploitation of synergies through adoption of close cycles. These systems also need to be socially acceptable, environment friendly and economically viable. Keeping in view the above aspects in mind the experiment was conducted with following objectives:

- ✰ To address critical constraints in small farm systems for overall improvement.
- ✰ To increase the profitability of households and livelihood security.

Methodology

The experiment was conducted at farmer's field in Rania and Ellenabad blocks of Sirsa district in six villages at 12 locations. The experiment was started during *Kharif* 2013. The details of the treatments are given as below:

Details of the Treatments

M_0: Cotton-wheat + Buffalo

M_1: M_0 + Critical constraints of crop components *i.e.* Improved variety and recommended fertilization

M_2: M_0 + Critical constraints of animal component *i.e* balanced feeding.

M_3: M_0 + Value addition of farm produce for more profit

M_4: M_0 + Nutritional kitchen gardening

Crop Yield and Economics

The data in Table 1 reveals that there was increase in grain yield with the application of zinc sulphate and potash as compared to farmers practices (T_0). The yield of cotton in improved practice was 2065, 2021, 1898 and 1995 kg/ha during 2013-14, 2014-15, 2015-16 and mean, respectively. The yield of wheat in improved practice was 4864, 4268, 5040 and 4724 kg/ha during 2013-14, 2014-15, 2015-16 and mean, respectively.

The gross returns and return over variable cost/net returns in improved practices was Rs. 173204, 143131, 175570 and 163968; 89029, 54700, 91036 and 78255 ha^{-1}, respectively during all the years of study, whereas, the gross returns and return over variable cost/net returns in farmer's practices was Rs.153831, 130580, 162946 and 149119; 73656, 51950, 83642 and 69749 ha^{-1}, respectively during all the years of study.

The mean benefit cost ratio was highest 2.25 in improved practice as compared to farmers practice *i.e.* 1.79.

Mineral Mixture as Input given to Farmer for IFS Experiment under OFR Sirsa.

OFR Farmer with his Buffalo.

OFR Farmer with her Buffalo.

OFR Farmer in his Kitchen Garden.

Table 1: Effect of Agronomic Management Practices on Yield and Economics in Cotton Wheat Cropping System (Average of 12 sites)

Treatment	*2013-14*		*2014-15*		*2015-16*		*Mean*	
	Farmer's Practice	*Improved Practice*	*Farmer's Practice*	*Improved Practice*	*Farmer's Practice*	*Improved Practice*	*Farmer's Practice*	*Improved Practice*
Cotton (Kg/ha)	1790	2065	1830	2021	1715	1898	1778	1995
Wheat (Kg/ha)	4480	4864	3932	4268	4790	5040	4401	4724
Gross returns (Rs/ha)	153831	173204	130580	143131	162946	175570	149119	163968
Return over variable cost/Return over variable cost/net return (Rs/ha)	73656	89029	51950	54700	83642	91036	69749	78255
B:C	1.91	2.05	2.51	2.62	2.05	2.08	2.16	2.25

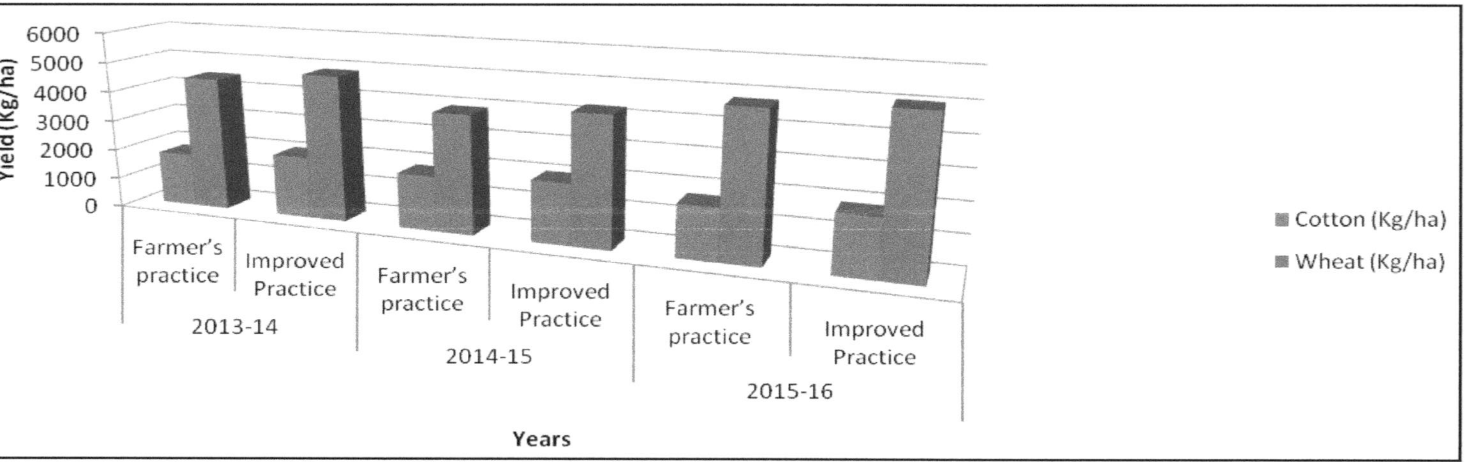

Figure 1: Effect of Agronomic Management Practices on Yield in Cotton Wheat Cropping System.

Table 2: Effect of Mineral Mixture and Nutritional Feed on Milk Yield and Economics of Milch Animal (Average of 12 sites)

Treatment	*2013-14*		*2014-15*		*2015-16*		*Mean*	
	Farmer's Practice	*Improved Practice*	*Farmer's Practice*	*Improved Practice*	*Farmer's Practice*	*Improved Practice*	*Farmer's Practice*	*Improved Practice*
Milk (Kg/animal/annum)	1522	1756	1698	1829	1826	1945	1682	1843
Gross returns (Rs/ha)	45660	52680	59430	64015	63910	68075	56333	61590
Return over variable cost/Return over variable cost/net return (Rs/ha)	15390	20870	27930	30705	27093	30098	23471	27224
B:C	1.51	1.65	2.13	2.08	1.74	1.79	1.79	1.84

Milk Yield and Economics

The milk yield was 1756, 1829, 1945 and 1843 kg per animal per annum with the use of mineral mixture and nutritional feed in the diet of animal during 2013-14, 2014-15, 2015-16 and mean, respectively (Table 2). The return over variable cost/ net returns in improved practice of feeding mineral mixture and nutritional feed to animal was Rs. 20870, 30705, 30098 and 27224 per animal per annum whereas return over variable cost/net return in farmer's practice was Rs. 15390, 27930, 27093 and 23471 per animal per annum during 2013-14, 2014-15, 2015-16 and mean, respectively. B: C under farmer's practice was 1.51, 2.13, 1.74, 1.79, whereas, under improved practice it was 1.65, 2.08, 1.79 and 1.84 during 2013-14, 2014-15, 2015-16 and mean, respectively.

Value Addition of Milk (Ghee making) and its Economics

The farmers were motivated to make the ghee from milk than sale of milk in the market and study was conducted to find out the additional profit due to value addition. The data in Table 3 shows that the making of ghee by the farmers has increased the average profitability by Rs. 8040, 7490 and 7250 over the sale of pure milk in the market during 2013-14, 2014-15 and 2015-16, respectively.

Table 3: Value Addition of Milk (Ghee making) and its Economics (Average of 12 sites)

Treatment (T_3)	*2013-14*	*2014-15*	*2015-16*
Quantity of raw material used (kg or no's or litres)	1160	598	643
Value of milk (Rs.)	35960	20357	22505
Production of ghee (kg)	80.0	46.0	49.6
Value of ghee (Rs.)	44000	27847	29755
Increase in profit in ghee making (Rs.)	8040	7490	7250

Chapter 10

Emerging Problem

New Emerging Problems and Constraints with the Farmers

1. Non-availability, poor and untimely availability of fertilizers
2. Global warming as a result of which unpredictable weather
3. Erratic and limited power supply
4. Depleting soil fertility and organic carbon
5. Weeds problem
6. Less response to nutrients
7. Brackish water
8. Price fluctuation
9. Marketing problems
10. Under utilization of farm waste and farm machinery
11. Non-availability of irrigation water at critical crop growth stages
12. Increasing indebtedness of the farmers

Chapter 11

Research Thrust

1. Nutrients status map to be prepared block wise.
2. Besides dairy component in IFS model, components like poultry, bee-keeping etc. should be introduced.
3. Proper economics of model should reach to the farmer level through agencies.
4. Arrangements are to be made for block wise irrigation according to crops proposed through irrigation societies.
5. Water economy has to be accomplished by suitable technology without adverse effect on yield.
6. Emphasis on inclusion of legumes and application of organic manure for sustainability.
7. Restriction on digging bore having brackish water.
8. Growing short duration varieties in lieu of long and medium duration varieties to have high cropping intensity.

Chapter 12

Research Publications

Research publications are important part of any scientific project. These publications are helpful in disseminating the scientific knowledge earned through conduct and compilation of research experiments. The important publications published by the scientist of the project are presented below:

A. Book

1. Question Bank on Agriculture. 2014 by R.K.Nanwal, Published by Prashant Book Agency, N. Delhi. ISBN: 978-81-922290-5-8.
2. Text book on farming systems 2016 by R.K.Nanwal and Pawan Kumar Published by Prashant Book Agency, N.Delhi. 2016. ISBN: 978-93-84502-03-4
3. Principles and Practices in dry land farming. 2016 By R.K.Nanwal and G. A. Rajanna ISBN : 878-93-85516-49-8. Published by NIPA, N.Delhi
4. Ebook. Growth, yield and quality of sorghum as influenced by nitrogen levels, 2014. Cleto Namboo and R.K.Nanwal, Lap Lambert Acad Pub, Germany ISBN no. 978-3-659-59248-5
5. Ebook. Production efficiency of moongbean-wheat cropping system, 2015. Neelam and R.K.Nanwal, Lap Lambert Acad Pub, Germany ISBN no. 978-3-659-61266-4
6. Samanvit krishi parnali- Paramprik kheti ka ek vikalp (In Hindi). 2016. By R.K.Nanwal *et al.*, To be published by Astral International Publishers, N.Delhi (In press)

B. Teaching Manuals

1. Rainfed Agriculture (2012) by RK Nanwal, Parveen Kumar, Pawan Kumar and AS Dhindwal

2. Cropping System and Sustainable Agriculture (2012) by Pawan Kumar, RK Nanwal, AS Dhindwal and S.K. Yadav.
3. Farming Systems and Sustainable Agriculture (2013) by R.K.Nanwal, Pawan Kumar and D.P. Nandal.
4. Dryland Farming and Watershed Management (2013) by RK Nanwal, Meena Sewhag, Parveen Kumar and AS Dhindwal.
5. Current Trends in Agronomy (2013) by R.K.Nanwal, A.K.Dhaka, Bhagat Singh and R.K.Pannu.
6. Integrated Farming Systems and Sustainable Agriculture (2014) by R.K.Nanwal, Pawan Kumar and Jagdev Singh.
7. Soil Conservation and Watershed Management (2014) by RK Nanwal, AK Dhaka, Bhagat Singh and Jagdev Singh.
8. Integrated Farming system (2014) by Pawan Kumar, RK Nanwal and Jagdev Singh.

C. Book Chapter

1. Yadav S K, Kumar Pawan, Kumar Manoj and Singh K. P. 2009. Viability of different cropping systems in semi-arid parts of Haryana. In: Diversification of Arid Farming Systems (Pratap Narain *et al.*: eds.) Arid Zone Research Association of India and Scientific Publishers (India), Jodhpur. pp148-152.
2. Rana, V.S.; Rathore, B.S.; Nanwal, R.K. and Kumar, Pawan. 2010. Influence of plant density and nutrient management on pearl millet in semi arid environment, In: Resource management towards sustainable agriculture and development (Eds: Behl *et al.*). Published by Agrobios (International), pp:49-63.
3. Kumar, Pawan; Yadav, S.K. and Kumar, Maonj. 2011. Efficient alternative cropping systems in trans gangetic plain zones of Haryana. In: Efficient Alternative Cropping Systems (Eds: Gangwar and Singh). Published by Project Directorate for Farming Systems Research, ICAR, Modipuram, pp:59-68.
4. Pawan Kumar, Manoj Kumar and S.K. Yadav. (2014) Enhancing pulses production: Haryana, In: Enhancing pulses production, by Gangwar and Singh, pp 141-160.
5. Kumar, Pawan; Sharma, Manoj Kumar; Yadav, S.K and Nanwal, R.K 2015. Organic nutrient management package for crop yield, soil fertility, nutrient uptake and crop quality in mungbean-wheat cropping system on a haplustalf of Hisar. In Mimiogram No.1/2015. Published by ICAR-Indian Institute of Farming System Research, Modipuram. pp. 163-183.
6. Kumar, Pawan; Sharma, Manoj Kumar; Nanwal, R.K and Yadav, S.K. 2015. Long term integrated nutrient management in pearl millet – wheat cropping system. Chapter in book entitled, "Long term integrated nutrient management in cereal based cropping systems" Published by ICAR-Indian Institute of Farming System Research, Modipuram. pp. 301-317.

D.Technology Bulletin

1. Yadav S.K. 2010. General Principles of Crop Production in Compendium on Commercial Agriculture, Published by Students Counseling and Placement Centre in Collaboration with Department of Agronomy, pp 1-10.
2. Kumar Pawan. 2010. Cropping Systems for Commercial Agriculture. In: Compendium on Commercial Agriculture, Published by Students Counseling and Placement Centre in Collaboration with Department of Agronomy, pp 181-188.
3. Kumar Pawan. 2011. Diversification in Agriculture. In: Compendium on Commercial Agriculture, Published by Students Counseling and Placement Centre in Collaboration with Department of Agronomy.
4. Yadav S K. 2011. Diversification in Agriculture: A General View. In: Compendium on Commercial Agriculture, Published by Students Counseling and Placement Centre in Collaboration with Department of Agronomy.
5. Farming Systems Research at CCS HAU, Hisar (2014) written by Drs. R.K.Nanwal; Pawan Kumar; Manoj Kumar; A.K. Mehta and Jagdev Singh Published by Deptt. of Agronomy.
6. Krishi Parnali Kisan Ke Khet Par Anusandhan (in Hindi) (2014) written by Drs. R.K.Nanwal; A.K. Mehta; Pawan Kumar; Manoj Kumar; and Jagdev Singh. Published by Deptt. of Agronomy
7. Samnvit Krishi Parnali Model (in Hindi) (2015) written by Drs. R.K.Nanwal; Pawan Kumar and Jagdev Singh. Published by Deptt. of Agronomy.
8. Phasal Vividhikaran Apnai Labh Payain (in Hindi) (2015) written by Drs. R.K.Nanwal; Pawan Kumar and Jagdev Singh. Published by Deptt. of Agronomy.

E. Research Papers

1. Bagla, Gaurav, Singh, Ishwar, Yadav, S. K. and Kumar, Pawan. 2007. Response of pearl millet to organic and inorganic sources of nutrition. *Natnl. J. Pl. Improv.* **9**(2): 99-102.
2. Jat, R.D. and Nanwal, R.K. 2013. Growth, nutrient uptake and profitability of Bt cotton (*Gossypium hirsutum*) influenced by spacing and nutrient levels. *Crop Res.* 45(1, 2 and 3):248-252.
3. Jat, R.D, Nanwal, R.K., Kumar, Pawan and Shivran, A.C. 2014. Productivity and nutrient uptake of Bt cotton (*Gossium hirsutum* L) under different spacing and nutrient levels. *J.Cotton Res. Dev.* 28(2):260-262.
4. Kumar Anil, Hooda R.S., Yadav H.P., Chugh L.K., Kumar Manoj and Gera Rajesh. 2009. Compensating nutrient requirement in pearl millet (*Pennisetum glaucum*)-wheat (*Triticum aestivum*) cropping system through manures and biofertilizers in semi-arid regions of Haryana. *Indian Journal of Agricultural Sciences* **79**(10): 767-71.

5. Kumar Pawan, Yadav S.K. and Kumar Manoj. 2008. Production potential and economics of different cropping systems in semi-arid zone of Haryana under irrigated situations. *Haryana J. Agron.* **24**(1 and 2): 59-61.

6. Kumar, Manoj, Kumar, Pawan, Yadav, S.K. and Pahuja, S. S. 2007. Crop management influence on nutrients uptake in pearl millet-wheat cropping system. Published in "Third National Symposium on Integrated Farming Systems and its Role Towards Livelihood Improvement" at ARS Jaipur, Oct. 26-28, 2007. pp 259-60.

7. Kumar, Manoj, Yadav, S.K., Kumar, Pawan and Rinwa, R.S. 2006. Nutrients uptake as influenced by different cropping systems in semi-arid tract. *Haryana J. hortic. Sci.* **35**(3 and 4): 303.

8. Kumar, Pawan and Yadav, Shri Krishan. 2013. Importance of integrated farming system management in seed production. In: How the farmers be a quality seed producer (Eds. Arya, Kumar Rajesh; Kumar, Suresh and Dahiya, K.K.) 36:133-135. (In Hindi)

9. Kumar, Pawan, Yadav, S.K., Kumar, Manoj and Chander, Ashok. 2013. Performance of field and vegetable based cropping systems in semi- arid Haryana, India. *Haryana J. hort. sci.*, 38 (1 and 2):101-103.

10. Kumar, Pawan and Nanwal, R.K. 2015. Pearl millet based cropping systems involving pulses, oilseeds and vegetables for attaining sustainability and economic viability in semi arid regions of India. *Indian Journal of Ecology*, **42**(1):148-151.

11. Kumar, Pawan, Nainwal, R.K. and Yadav, S.K. 2006. Economics of pearl millet-wheat Cropping system as influenced by organic and inorganic sources of nutrients. *Haryana Agric. Univ. J. Res.* 36: 13-16.

12. Kumar, Pawan; Kumar, Manoj, Yadav, S.K. and Nanwal, R.K. 2012. Effect of organic farming on dry fodder yield, grain yield, return over variable cost/net returns and soil fertility in moongbean-wheat (tall) production system. *Forage Res.*, 38(3): 177-181.

13. Kumar, Pawan, Kumar, Manoj, Yadav, S.K. and Nanwal, R.K. 2014. Agronomic management for sustainable pearl millet-wheat cropping systems in sandy loam soils. *Ann. Agri. Res. New Series* 35 (3): 304-310.

14. Kumar, Pawan, Nanwalj, Rajender, Kumar and Satyajeet. 2014. *Rabi* mein tilhan utpadan (in Hindi). *Khad Patrika* 55(9):12-15.

15. Kumar, Pawan, Yadav, S.K., Kumar, Manoj, Rinwa, R.S. and Singh, K.P. 2007. Breaking yield barriers in pearl millet (*Pennisetum glaucum*)-wheat (*Triticum aestivum*) cropping system through agronomic operations in semi-arid Haryana. *Indian J. Agric. Sci.* **77**: 479-82.

16. Kumar, Pawan, Yadav, S.K., Kumar, Manoj and Hasija, R.C. 2014. Cropping systems involving pulses, oilseeds and vegetable for livelihood improvement of farmers in semi-arid region. *Annals of Biology* 30(1) : 123-126.

17. Kumar, Pawan, Yadav, S.K., Kumar, Manoj, Kumar, Suresh and Hasija, R.C. 2011. Effect of FYM and higher plant population on yield and water use efficiency in pearl millet-wheat cropping system. *Environ. and Ecol.*, 20: 190-93.

18. Kumar, Pawan, Yadav, S.K., Kumar, Suresh and Kumar, Manoj. 2009. Sustaining yield of pearl millet-wheat cropping system in semi arid conditions. *Haryana J. Agron.* **25**(1 and 2): 1-3.

19. Namoobe, Cleto, Nanwal, R.K. and Kumar, Pawan. 2014. Productivity and economics of sorghum varieties for grain as influenced by nitrogen levels in sandy loam soil. *International J. Natural Sciences Research* 2(1): 5-11.

20. Namoobe, Cleto, Nanwal, R.K, Kumar, Pawan and Grewal, K.S. 2012. Nutrient uptake by grain sorghum as influenced by cultivars and nitrogen fertilization in sandy loam soil. *Forage Res.* 38(3): 129-132.

21. Nanwal, R.K. 2016. Diversification of rice-wheat cropping system to improve soil fertility, sustainable productivity and economics in IGP. Lead paper presented in Indian Ecological Society: International Conference on Natural Resource Management- Ecological Perceptives held on Feb 18-20, 2016 and published in Vol I pp.16.

22. Nanwal,R.K., Pawan Kumar Kumar, Manoj and Singh, Jagdev. 2015. Diversified farming system model: An agro-ecological cropping based alternative for enhancement of rural livelihood. Abs in National seminar on IFS for sustainable agriculture and enhancement of rural livelihood Dec. 5-6, 2015 at ICAR-NAARAM, Hyderabad.

23. Neelam, Nanwal, R.K. and Kumar, Pawan. 2013. Productivity and quality of mungbean-wheat cropping system as influenced by organic and inorganic sources of nutrients in semi arid environment. *Res. on Crops* 14(3): 786-791.

24. Neelam, Nanwal, R.K. and Kumar, Pawan. 2014. Nutrient management in pulse-wheat cropping system: A Review. *Annals of Agri. Bio. Res.* 19(1):61-66.

25. Neelam, Nanwal, R.K. and Kumar, Pawan. 2015. Effects of organic and inorganic sources of nutrients on productivity and profitability of mungbean-wheat cropping system. *Legume Res.* 38 (4):509-512.

26. Neelam, Nanwal, R.K. Kumar, Pawan and Grewal, K.S. 2015. Productivity and nutrient uptake studies in mungbean-wheat cropping system as influenced by integrated nutrient management in semi arid conditions. *Environment and Ecology*. 33 (1):110-114.

27. Parihar, M.D., Nanwal, R.K., Kumar, Pawan., Kumar, Satish., Singh, A.K., Chaudhary, V., Parmer, H and Jat, M.L.2015. Effect of tillage practices and cropping systems on growth and yield of maize grown in sequence with wheat and chickpea. *Ann. Agric. Res. New Series*. 36(2):177-183.

28. Pawan Kumar, Yadav, S.K, Kumar, Manoj and Nanwal, R.K. 2012. Effect of integrated nutrient management on crop productivity and soil health under cereal-cereal cropping sequence. *Haryana J.Agron.* **28**(1 and 2):50-52.

29. Rajanna, G.A., Dhindwal, A.S. and Nanwal, R.K. 2016. Effect of irrigation schedules on plant water relations, root studies, grain yield and water productivity of wheat under various crop establishment techniques. International J. Plant Production. Accepted vide no. IJPP-1606-1342.

30. Rawal S., Mehta A.K., Thakral S.K., Mahesh Kumar. 2015. Effect of nitrogen and phosphorus levels on growth, yield attributes and yields of Bt cotton. *J. Cotton Res.Dev.* 29 (1)76-78.

31. Rawal S., Mehta A.K., Thakral S.K. and Kumar M. (2015).Effect of Nitrogen and Phosphorus levels on Nutrient content,uptake and yield of Bt cotton under irrigate conditions. Man made Textiles in India. Oct. 2015 381-384.

32. Sangwan, Naresh, Amandeep, Singh, Ran, Yadav, S.K. and Kumar, Manoj. 2007. Response of rice (*Oryza sativa*) to NPK levels on yield and yield attributes and nutrients uptake in rice-wheat cropping system at farmers' field. *Haryana J. Agron.* **23**(1 and 2): 39-42.

33. Singh, Kanwar, Malik, R.K., Yadav, S.K., Yadav, Ashok and Singh, Sher. 2006. Effect of irrigation and chlorsulfuron on wheat (*Triticum aestivum* L.). *Haryana agric. Univ. J. Res.*, 36:41-44.

34. Singh, Kanwar, Malik, R.K., Yadav, S.K.,Yadav, A, Singh, Sher and Sangwan, N.K. 2006. Effect of irrigation levels and chlorsulfuron doses on nutrient content and uptake of wheat (*Triticum aestivum* L.). *Annals Agri-Bio Res.*, 11 : 141-145.

35. Yadav, Rekha, Nanwal, R.K. and Kumar, Anil. 2014. Effect of potash, zinc and biofertilizer application on productivity and quality of pearl millet. *Environment and ecology* 32(3): 980-983.

36. Yadav, Rekha; Nanwal, R.K. and Kumar, Anil. 2014. Effect of potash, zinc and biofertilizer application on yield and nutrient uptake in pearl millet. *Annals of biology* 30(4): 645-648.

37. Yadav, Rekha; Nanwal, R.K. and Kumar, Anil. 2014. Maximization in pearl millet productivity and profitability through potash, zinc and biofertilizer application. *Environment and ecology* 32(4): 1661-1664.

38. Yadav, S.K., Kumar, Pawan and Kumar, Manoj. 2012. Cropping systems for higher productivity in semi-arid low rainfall areas of Haryana. Paper presented and abstract published in Extended Summaries, vol 2 of 3rd International Agronomy Congress on Agriculture Diversification, Climate change management and Livelihood, held at IARI New Delhi from Nov. 26 to 30,2012, organized by Indian Society of Agronomy and ICAR.

39. Yadav, S.K., Kumar, Pawan, Nanwal, R.K., Kumar, Manoj and Rana, V.S. 2012. Diversification in agriculture- issues and actions. In: Crop Science and Technology for Food Security, Bioenergy and Sustainability (Eds. R.K. Behl,L. Bona, J. Pauk, W. Merbach and A. Veha)pp 345-350, Published by Agrobios (International), Jodhpur

40. Yadav, S.K.; Kumar, Pawan and Kumar, Manoj 2010. Possible alternative cropping systems for farmers of semi arid Haryana-India. *J. Farming Systems Res. and Development*. 16:36-42.

F. Paper Presented in Symposium/Seminar

1. Kumar, Pawan and Yadav, S.K. 2006. Sustainable crop production through the use of Various organic sources in association with chemical fertilizers in pearl millet-wheat cropping system. Abstract published in conference on "Natural Resources management for Sustainable Development in Western India (NRMSD-2006)", held at Pune from 11-13 October, 2006.
2. Kumar, Pawan, and Yadav, S.K. 2006 Enhancing yield potential with the application of FYM and higher seed rate in pearl millet – wheat cropping system. Extended summary Published in Golden Jublee National Symposium on 'Conservation Agriculture and Environment' organized by Indian Society of agronomy at BHU, Varanasi from Oct. 26-28, 2006.
3. Chauhan, D. R.; Om, Hari; Rinwa, R. S. and Yadav, S. K. 2007. Evaluation of frontline demonstration on pearl millet-raya cropping system in western Haryana. Published in "Third National Symposium on Integrated Farming Systems and its Role Towards Livelihood Improvement" at ARS Jaipur,
4. Hari Om; Rinwa, R. S.; Chauhan, D. R.; Yadav, S. K. and Chauhan, R. S. 2007. Effect of constraint management in rice/pearl millet/cotton-wheat cropping sequences. Published in "Third National Symposium on Integrated Farming Systems and its Role Towards Livelihood Improvement" at ARS Jaipur, Oct. 26-28, 2007. pp 170-71.
5. Kadian, V. S.; Kumar, Pawan; Yadav, S. K. and Pahuja, S. S. 2007. response of cotton (*desi*)-wheat(*desi*) cropping system to various organic sources of nutrients. Published in "Third National Symposium on Integrated Farming Systems and its Role Towards Livelihood Improvement" at ARS Jaipur, Oct. 26-28, 2007. pp 294-95.
6. Rinwa, R. S.; Hari Om; Chauhan, D. R.; Kumar, Manoj and Yadav, S. K. 2007. Effect of nutrient management on productivity, uptake and economics in cotton (*Gossypium hirsutum*)-wheat(*Triticum aestivum*) system. Published in "Third National Symposium on Integrated Farming Systems and its Role Towards Livelihood Improvement" at ARS Jaipur, Oct. 26-28, 2007. pp 235-36.
7. Rinwa, R. S.; Hari Om; Chauhan, D. R.; Kumar, Manoj and Yadav, S. K. 2007. Impact of balanced fertilizer in pearl millet (*Pennisetum glaucum*)- wheat(*Triticum aestivum*) cropping system. Published in "Third National Symposium on Integrated Farming Systems and its Role Towards Livelihood Improvement" at ARS Jaipur, Oct. 26-28, 2007. pp 261-62.
8. Rinwa, R. S.; Hari Om; Chauhan, D. R.; Kumar, Manoj and Yadav, S. K. 2007. Influence of balanced fertilization on yield, nutrient uptake and economics in rice (*Oryza sativa*)-wheat(*Triticum aestivum*) cropping system. Published

in "Third National Symposium on Integrated Farming Systems and its Role Towards Livelihood Improvement" at ARS Jaipur, Oct. 26-28, 2007. pp 263-64.

9. Kumar, Manoj; Kumar, Pawan; Yadav, S. K. and Pahuja, S. S. 2007. Crop management influence on nutrients uptake in pearl millet-wheat cropping system. Published in "Third National Symposium on Integrated Farming Systems and its Role Towards Livelihood Improvement" at ARS Jaipur, Oct. 26-28, 2007. pp 259-60.

10. Kumar, Pawan; Pahuja, S. S.; Yadav, S. K. and Hari Om. 2007. Mungbean-wheat (*desi*) cropping system response to different organic sources. Published in "Third National Symposium on Integrated Farming Systems and its Role Towards Livelihood Improvement" at ARS Jaipur, Oct. 26-28, 2007. pp 296-97.

11. Yadav, S. K.; Kumar, Pawan; Kumar, Manoj and Pahuja, S. S. 2007. Effect of different organic sources of nutrients on mungbean-*desi* wheat cropping system in semi-arid Haryana. Extended Summary published in "International Symposium on Organic Farming and Renewable Sources of Energy for Sustainable Agriculture", held at MPUAT Udaipur, Nov. 19-21, 2007. pp 146-147.

12. Yadav, S. K.; Kumar, Pawan; Kumar, Manoj and Pahuja, S. S. 2007. Yield and economics of various cropping systems in semi-arid Haryana. Published in "Third National Symposium on Integrated Farming Systems and its Role Towards Livelihood Improvement" at ARS Jaipur, Oct. 26-28, 2007. pp 149-50.

13. Yadav, S. K.; Kumar, Pawan; Kumar, Manoj 2008. Productivity and net return of various cropping systems in semi-arid Haryana. Paper presented and abstract published in *National Seminar on Recent Trends in Development of Spices and Aromatic Plants,* Sept. 10-12, 2008, held at CCS Haryana Agricultural University, Hisar. p. 30.

14. Kumar, Pawan; Yadav, S. K.; Kumar, Manoj 2008. Organic cultivation of mungbean (*Vigna radiata*)-*desi* wheat (*Triticum aestivum*) cropping system. Paper presented and abstract published in *National Symposium on New Paradigm in Agronomic Research,* organized by Indian Society of Agronomy, Nov. 19-21, 2008 held at Navsari Agricultural University, Navsari, Gujarat. p.516.

15. Kumar, Pawan; Yadav, S. K.; Kumar, Manoj 2009. Productivity in different cropping systems grown in semi-arid Haryana. Paper presented and abstract published in *"Forage Symposium – 2009 Emerging Trends in Forage Research and Livestock production"*, Feb. 16-17, 2009 held at CAZRI, RRS Jaisalmer.

16. Yadav S K, Kumar Pawan and Kumar Manoj. 2009. Management of pearlimillet-wheat cropping system for sustainable yield. Abstract published in: National Conference on Frontiers in Plant Physiology towards sustainable Agriculture, November 5-7, 2009, Jorhat, Assam.

17. Yadav, S.K; Kumar, Pawan and. Kumar, Manoj. 2010. Integrated nutrient management for enhanced productivity and profitability of pearl millet-wheat cropping system. Abst. Published in : National Symposium on Emerging Trends in Agricultural Research, 11-12 Sept, 2010 at PDFSR, Modipuram, Meerut, pp:33.

18. Hari Om; Yadav, S.K.; Kumar, Manoj; Kumar, Pawan and Dilbagh Singh. 2010. Impact of nutrient management on productivity and economics in rice-wheat system. Abst. Published in : National Symposium on Emerging Trends in Agricultural Research, 11-12 Sept, 2010 at PDFSR, Modipuram, Meerut, pp:31.

19. Hari Om; Yadav, S.K.; Singh, Dilbagh; Saini, R.S. and Pannu, R.S. 2010. Diversification of rice-wheat system in north-western Indian conditions. Abst. Published in : National Symposium on Emerging Trends in Agricultural Research, 11-12 Sept, 2010 at PDFSR, Modipuram, Meerut, pp:29-30.

20. Kumar, Manoj; Hari Om; Yadav, S.K.; Singh, Dilbagh and Lathwal, O.P. 2010. Response and uptake of nutrients under different nutrient combinations in rice-wheat system. Abst. Published in : National Symposium on Emerging Trends in Agricultural Research, 11-12 Sept, 2010 at PDFSR, Modipuram, Meerut, pp:30-31.

21. Kumar, Manoj; Yadav, S.K. and Kumar, Pawan. 2010. Cropping systems influence on nutrients utilization in semi arid Haryana. Abst. Published in : Symposium on Emerging Trends in Agricultural Research, 11-12 Sept, 2010 at PDFSR, Modipuram, Meerut, pp:32.

22. Kumar, Pawan; Yadav, S.K. and Kumar, Manoj. 2010. Various cropping systems to minimize risk in agriculture. Abst. Published in : Symposium on Emerging Trends in Agricultural Research, 11-12 Sept, 2010 at PDFSR, Modipuram, Meerut, pp:33.

23. Kumar, Pawan; Kadian, V.S.; Yadav, S.K. and Kumar, Manoj. 2011. Effect of various organic sources of nutrients on yield of cotton(*Desi*)-wheat cropping system. Abst. Published in National Symposium-cum-Brain Storming Workshop on Organic Agriculture, held at CSK HPKVV Palampur during 19-20 April, 2011, pp:81-82.

24. Kumar, Pawan; Nanwal,R.K. and Yadav, S.K. 2014. Integrated Cropping System Model for Small and Marginal Farming Situations of Haryana. Paper presented in National Seminar on Reorientation of Agricultural Research to Ensure National Food Security, Jan., 6-7, 2014 at CCS HAU, Hisar. Pp.164.

25. Nanwal,R.K. and Pawan Kumar. 2014 Integrated Farming System Model for irrigated as well as rainfed farming situations for Haryana. Paper presented in National Seminar on Reorientation of Agricultural Research to Ensure National Food Security, Jan., 6-7, 2014 at CCS HAU, Hisar. Pp.167.

26. Sharma, Manoj, Kumar; Yadav, S.K; Pawan Kumar; and Nanwal, R.K. 2014. Integrated Nutrient Management for better soil fertility and nutrient uptake. Paper presented in National Seminar on Reorientation of Agricultural Research to Ensure National Food Security, Jan., 6-7, 2014 at CCS HAU, Hisar. Pp.142.

27. Kumar, Pawan; Manoj, Kumar; Yadav, S.K. and Nanwal,R.K. 2014. Integrated nutrient management in pearl millet-wheat cropping system for sustained production. In : 101st Indian Science Congress Vol. II pp. 69-79 held at Jammu university, Jammu (3-7 Feb, 2014.

28. RK Nanwal attended and presented research work of Hisar centre at the Group meeting of AICRP on IFS at AAU, Jorhat during 16-18, December 2015.

29. Yadav, N.K., Nirania, K.S., Bhattoo, M.S. and Mehta, A. K. (2015). Evaluation of American cotton (*Gossypium hirsutum* L.) genotypes for resistance to Cotton Leaf Curl Disease (CLCuD) Paper presented in National symposium on "Future Technologies: Indian cotton in the Next Decade" held at A.N. University, Gunture (A.P.) -522510 on 17-19 Dec,2015 Ab.pp 85-86.

30. Nirania K.S.,Yadav, N.K. and Mehta, A. K. (2015). Heterosis and combining ability in American cotton (*Gossypium hirsutum* L.) Paper presented in National symposium on "Future Technologies: Indian cotton in the Next Decade" held at A.N. University, Gunture (A.P.) -522510 on 17-19 Dec,2015 Ab.pp 27.

31. Sriharsha V.P.,Mehta A.K. and Rawal Sandeep (2015). Foliar feeding of fertilizers on Bt cotton (Gossypium hirsutum L.) Paper presented in 102nd Indian Science Congress held at University of Mumbai,Mumbai on 03-07 Jan.,2015 pp 136.

32. RK Nanwal attended and presented lead paper in International Conference on Natural Resource Management held at SKUAST, Jammu during Feb 18-20, 2016.

G. Popular Articles

1. Kumar Pawan and Yadav S K. 2009. Mulching ka mahatav. *Haryana Kheti* **42**(6): 4.

2. Kumar Pawan, Yadav S K and Singh Attar. 2009. Phasal avshesh "Mahatvpuran avum rakshniy. *Haryana Kheti* **42**(4): 29.

3. Kumar, Pawan; Rajender Kumar Nanwal and Satyajeet. 2014. *Rabi* mein tilhan utpadan. Khad Patrika 55(9):12-15.

4. Kumar, Pawan; Yadav, S. K. and Dhukia, R. S. 2006. Urvarko ka uchit paryog. Haryana Kheti (Hindi). 12: 6.

5. Kumar, Pawan; Yadav, S. K. and Nanwal, R. K. 2006. Bajra ki unnat kheti. Haryana Kheti (Hindi).

6. Kumar, Pawan; Yadav. S. K. and Yadav, Ashok 2006. *Kharif* faslo ki dekhbhal. Haryana Kheti (Hindi). 8: 4-5.

7. Mehta A. K.(2015). 'Success of Marginal Farmer with IFS approach.' in Farming Systems Research Success stories (series 1) ICAR-IIFSR pp 54-55.

8. Neelam, Rajender Kumar Nanwal and Pawan Kumar. 2014. Kanchuwa Ki Khad Ka Krishi Mein Mahatav (In Hindi). Haryana Kheti, Dec, 2014: 5.

9. Neelam, Rajender Kumar Nanwal and Pawan Kumar. 2014. Kanchuwa Khad (Vermicompost) Tayar Karene Ki Vidhi (In Hindi). Haryana Kheti, Dec, 2014: 27-28.

10. Neelam; Pawan Kumar and Rajender Kumar Nanwal. 2013. Tikao utpadan ke liye kargar hai javik kheti. Haryana Sanwad 45(12): 133-137.

11. Satyajit; Kumar, Pawan and Nanwal, R.K. 2011. Jaivik Khad:Aadhunik Krishi Mein Tikau Utpadan Ke Liye nai Dishain. Khad Patrika, Feb, 2011.
12. Yadav N K, M.S. Bhatto,K S Nirania and Anil Kumar Mehta. 2015. Desi Kapas Ki Unnat Kheti. Krishi Sanwad May 2015 pp. 22-23.
13. Yadav, S. K.; Kumar, Pawan and Malik, R. K. 2006. Jaivik utpado ka parmanikaran. Haryana Kheti (Hindi). 11: 27.

H. Symposia/Conferences/Trainings Attended

1. Attended, presented Hisar work *and* planned activities for the next year in XXX Workshop of AICRP on IFS from 16-19 Nov. 2012 at ICAR Research complex for Goa.
2. Attended 3rd International Agronomy Congress on Agriculture Diversification, Climate change management and Livelihood, held at IARI New Delhi from Nov. 26 to 30, 2012, organized by Indian Society of Agronomy and ICAR.
3. Attended and presented Hisar centre work in the Group Meeting of AICRP on IFS from 2-4, Dec. 2013 at ICAR Research complex at Umiam (Shillong). (Dr. R.K. Nanwal).
4. Attended National Seminar on Reorientation of agricultural research to ensure national food security held from Jan. 6-7, 2014 at CCS HAU, Hisar. (Team of IFS)
5. Organised and attended Regional Group Meeting of AICRP on IFS organized in collaboration with IIFSR, Modipuram from Aug 11-13, 2014 at CCSHAU, Hisar.
6. Attended and presented research work of Hisar centre at the Biennual Workshop of AICRP on IFS at TNAU, Coimbtore during 22-24, December 2014.
7. Attended technical programme Review Workshop of AICRP on IFS at IIFSR, Modipuram from 18-19, May 2015.
8. Attended and presented research work of Hisar centre at the Group meeting of AICRP on IFS at AAU, Jorhat during 16-18, December 2015.
9. Attended and presented lead paper in International Conference on Natural Resource Management held at SKUAST, Jammu during Feb 18-20, 2016.

Chapter 13

Visits to the Experiments

Various dignitaries, scientist, students and farmers are regularly visiting this project. The most important among these is the farmer who carries the technology generated from this project in field and lab to his land. This contributes in enhancing the productivity and profitability of important cropping and farming systems adopted in the state. Important visitors visited the project in the recent past is given below:

Year 2007-08

- ☆ Director of Research, CCS HAU, Hisar visited the cropping systems research experiments many times in *Kharif* and *Rabi* seasons.
- ☆ Director of Extension Education CCS HAU, Hisar visited the cropping systems research experiments.
- ☆ A team of weed control scientists visited the cropping systems research experiments.
- ☆ Dr. H.D. Yadav, Senior Co-ordinator KVK Mohindergarh, Visited the cropping systems research experiments.
- ☆ Prof. and Head, Department of Soil Science visited the cropping systems research.
- ☆ A group of students of North Eastern states visited the cropping systems research experiments.
- ☆ Dr. A.C. Yadav, Sr. Scientist (Vegetable Scientist) visited the farming systems research experiments.
- ☆ Prof. and Head, Department of Agronomy regularly visited the cropping systems research and ECF experiments.
- ☆ Dean, College of Agriculture visited the cropping systems research experiments.

- ☆ Farmers from different parts of Haryana visited the farming systems research experiments.
- ☆ Dr S.S. Pal, Programme Facilitator, PDFSR visited the cropping system research and ECF experiments on Feb., 27 and 28, 2008.

Year 2008-09

- ☆ Director of Research, CCS HAU, Hisar visited the cropping systems research experiments many times in *Kharif* and *Rabi* seasons.
- ☆ Director of Extension Education CCS HAU, Hisar visited the cropping systems research experiments.
- ☆ Professor David Coventry from Adelaide University, Australia with ACIAR Project review team visited the cropping systems research experiments.
- ☆ Delegates of world congress on conservation agriculture visited the experiments of cropping system research.
- ☆ Dr, N.C. Turner, Sr. Chief Researcher from Australia visited the cropping system Research experiments.
- ☆ A team of scientists from CCSRI, Karnal visited the farming system experiments.

Year 2009-10

- ☆ Director of Research, CCS HAU, Hisar visited the cropping systems research experiments many times in *Kharif* and *Rabi* seasons.
- ☆ Director of Extension Education CCS HAU, Hisar visited the cropping systems research experiments.
- ☆ Dr. A.C. Yadav, Sr. Scientist (Vegetable Scientist) visited the farming systems research experiments.
- ☆ Prof. and Head, Department of Agronomy regularly visites the cropping systems research and ECF experiments.
- ☆ Dean, College of Agriculture visited the cropping systems research experiments.
- ☆ Farmers from different parts of Haryana visited the farming systems research experiments.

Year 2010-11

- ☆ Director of Research, CCS HAU, Hisar visited the cropping systems research experiments many times in *Kharif* and *Rabi* seasons.
- ☆ Director of Extension Education CCS HAU, Hisar visited the cropping systems research experiments.
- ☆ Dr. A.C. Yadav, Sr. Scientist (Vegetable Scientist) visited the farming systems research experiments.

- ☆ PDFSR Modipuram team of Dr. Kamta Prasad, Programme Facilitator (CU) and Dr. J.P.Singh, Programme Facilitator (IFS), visited Hisar centre on 6th and 7th June, 2011.

Year 2011-12

- ☆ Director of Research, CCS HAU, Hisar visited the cropping systems research experiments many times in *Kharif* and *Rabi* seasons.
- ☆ Director of Extension Education CCS HAU, Hisar visited the cropping systems research experiments.
- ☆ Dr. A.C. Yadav, Sr. Scientist (Vegetable Scientist) visited the farming systems research experiments.
- ☆ Prof. and Head, Department of Agronomy regularly visites the cropping systems research and ECF experiments.
- ☆ Dean, College of Agriculture visited the cropping systems research experiments.
- ☆ Farmers from different parts of Haryana visited the farming systems research experiments.

Year 2012-13

- ☆ Director of Research, CCS HAU, Hisar visited the cropping systems research experiments many times in *Kharif* and *Rabi* seasons.
- ☆ Director of Extension Education CCS HAU, Hisar visited the cropping systems research experiments.
- ☆ Farmers from different parts of Haryana visited the farming systems research experiments.
- ☆ Dr. A.C. Yadav, Sr. Scientist (Vegetable Scientist) and Dr. Rakesh Kumar, Sr. Agronomist (Pulses) visited the farming systems research experiments.
- ☆ Prof. and Head, Department of Agronomy regularly visites the cropping systems research and ECF experiments.
- ☆ Dean, College of Agriculture visited the cropping systems research experiments.
- ☆ Undergraduate and post-graduate students of Agronomy courses *viz.* Agron. 305, Agron. 511 and Agron. 607 visited the research area of IFS scheme for practical purpose.

Year2013-14

- ☆ Dr. Kamta Prasad and Dr. Ravishanker visited IFS experiments at Hisar from 6-7, Jan. 2014.
- ☆ Dr. J.P. Singh and other members of monitoring team visited IFS experiments at Hisar and Sirsa in September 2013.

- Director of Research, Additional Director of Research and Project Director, CCS HAU, Hisar visited the farming systems research experiments in *Kharif* and *Rabi* seasons.
- Dr. R.K. Sharma along with other scientist from CIRB Hisar regularly visiting the dairy unit of the IFS model
- Farmers from different parts of Haryana visited the farming systems research experiments. The farmers were apprised of the IFS model during annual kisan mela of the University.
- Dr. A.C. Yadav, Sr. Scientist (Vegetable Scientist), Dr. D.S. Dahiya and Dr. Rakesh Kumar, Sr. Agronomist (Pulses) visited the farming systems research experiments.
- Prof. and Head, Department of Agronomy regularly visited the farming systems research experiments.
- Undergraduate and post-graduate students of Agronomy (specialized course on farming system) courses *viz.*, Agron. 305, Agron. 403, Agron. 511 and Agron. 607 visited the research area of IFS scheme for practical purpose especially IFS model.

Year 2014-15

- Dr Subash (Principal scientist) along with SRF from IIFSR, Modipuram visited the farming systems research experiments on 19 Dec., 2014.
- Dr A.R. Sharma Director from IIWR, Jabalpur visited the farming systems research experiments on 9 Feb., 2015.
- Dr. Kamta Prasad and Dr. Daleep Kachroo visited IFS experiments as member of the monitoring team at Hisar for on station and Sirsa for on farm experiments from 17-18, April 2014.
- Director of Research, Additional Director of Research and Project Director, CCS HAU, Hisar visited the farming systems research experiments in *Kharif* and *Rabi* seasons.
- Dr. R.K. Sharma along with other scientist from CIRB Hisar visited the dairy unit of the IFS model
- Farmers from different parts of Haryana visited the farming systems research experiments. The farmers were apprised of the IFS model during annual kisan mela of the University.
- Dr. A.C. Yadav, Sr. Scientist (Vegetable Scientist), Dr. D.S. Dahiya and Dr. Surjit Kumar, P.I. AICRP on Mushroom visited the farming systems research experiments.
- Prof. and Head, Department of Agronomy regularly visits the farming systems research experiments.

Year 2015-16

- ☆ Yog Rishi Baba Ramdev visited along with Dr KS Khokar, Vice Chancellor of the university to farming systems model on 24.10.2015.
- ☆ 6 IAS officers visited the farming systems research experiments on 02.12.2015.
- ☆ Nepal delegation visited IFS experiments on11.12.2015.
- ☆ IARI Principal Scientist Dr UK Behera visit IFS experiments on 07.01.2016.
- ☆ 20 farmers along with officers from Dept. of agriculture, Karnal visited IFS experiments on 27.02.2016.
- ☆ 7 members team from ministry of agriculture Govt. of Swaziland visited IFS experiments on 08.03.2016.
- ☆ Scientist with 6 farmers from AAU, Gujrat visited IFS model on 17.03.2016.
- ☆ Dr. J.S. Sandhu, DDG ICAR, Dr. O.P. Yadav, Director CAZRI along with VC, DR, Director HRM visited IFS model on 20.03.2016.
- ☆ Dr. V.P. Sharma, Director, NRC Mushroom visited IFS model on 04.07.2016.
- ☆ Dr K.P Singh, Vice Chancellor, CCSHAU, Hisar visited the farming systems model on 16.08.2016.
- ☆ Director of Research, Additional Director of Research and Project Director, CCS HAU, Hisar visited the farming systems research experiments in *Kharif* and *Rabi* seasons.
- ☆ Dr. R.K. Sharma along with other scientist from CIRB Hisar and Scientist from LUVAS namely Dr. Pandey and Dr. Sandeep Parihar regularly visited the dairy unit of the IFS model
- ☆ Farmers from different parts of Haryana visited the farming systems research experiments. The farmers were apprised of the IFS model during annual kisan mela of the University.
- ☆ Dr. Surjit Kumar, P.I. AICRP on Mushroom, Dr. D.S. Dahiya Professor (Horticulture), Dr. Dalip Bishnoi, Scientist (Agriculture Economics) and Dr. Makhan Lal, Scientist (Vegetable Scientist), visited the farming systems research experiments.
- ☆ Prof. and Head, Department of Agronomy regularly visiting the farming systems research experiments.

Monitoring Team from ICAR-Indian Institute of Farming System Research, Modipuram Visited the IFS Model.

Programme Facilitator from PDFSR at the IFS on Station Experiments during RARF (Jan 6-7, 2014).

Dr. J.S. Sandhu, DDG ICAR, and Dr. K.S. Khokar Visited IFS Model on 20.03.2016.

Yog Rishi Baba Ramdev and Dr. K.S. Khokar, Visited the Farming Systems Model on 24.10.2015.

Dr. K.P. Singh Visited the Integrated Farming System Model.

IAS Probation Officers Visit to IFS Model.

IAS Probation Officers Visit to IFS Model.

Nepal Delegates Visited IFS Model.

7 Members Team from Ministry of Agriculture, Govt. of Swaziland Visited IFS Model.

Dr. V.P. Sharma, Director, NRC on Mushroom Visited IFS Model.

Progressive Farmers from different Parts of Haryana Visited the Farming System Model.

References

Annonymous. 2009. http://agricoop.nic.in/Inm/cfinm9210.pdf

Annonymous. 2013. http://www.topnews.in/food-grain-production-2012-13.

Antil, R.S. and Mandeep, S. 2007. Effect of organic manures and fertilizers on organic matter and nutrients status of the soil. *Arch. Agron. Soil Sci.* **53**:519-528.

Antil, R.S. and Narwal, R.P. 2007. Role of integrated nutrient management for sustainable soil health and crop productivity under various cropping systems. *Indian J. Fert.* **3**: 111-121.

Axford, D. W. E. McDermott, E.E. and Redman, D.G. 1979. Note on the Sodium Dodecyl Sulphate Test of Bread Making Quality; Comparison with Pelshenke and Zeleny test. *Cereal Chemistry*, **56**: 582-584

Behera, U.K., Sharma, A.R. and Pandey, H.N. 2007. Sustaining productivity of wheat-soybean cropping system through integrated nutrient management practices on the vertisols of central india. *Plant and soil.*, **297** (1/2): 185-199.

Bhattacharya, R., Kundu, S., Ved Prakash and Gupta, H.S. 2008. Sustainability under combined application of mineral and organic fertilizers in a rainfed soybean-wheat system system of the Indian Himalayas. *Eur.J. Agron.*, **28**(1): 33-46.

Boyle, S.A.; Yarwood, R.R.; Bottomley, P.J. and Myrold, D.D. 2008. Bacterial and fungal contribution to soil nitrogen cycling under Douglas fir and red alder at two sites in Oregon. *Soil Bio. Biochem.* **40**: 443-451.

Cassman, K.G. and Pingali, P.L. 1993. Extrapolating trends from long term experiments to farmers fields: the case of irrigated rice systems in Asia. In: Proceedings of the Working Conference on Measuring Sustainability Using Long Term Experiments. Rothamsted Experimental Station, 28.30 April 1993, funded by the Agricultural Science Division, The Rockfeller Foundation.

Chakarborti, M. and Singh, N.P. 2004. Bio-Compost: a novel input to the organic farming. *Agrobios Newsletter*, **2**(8): 14-15.

Clark, M.S., Horwath, W.R., Shennan, C. and Scow, K.M. 1998. Changes in soil chemical properties resulting from organic and low-input farming practices, *Agronomy Journal*, vol. **90**, no. 5, pp. 662-671.

Fisher, R.A. 1958. *Statistical Methods for Research Workers.* Ed. 43: Oliver and Boyed, London.

Gangaiah, B., Ahlawat, I.P.S. and Shivkumar, B.G. 2012. Crop rotation and residue recycling effects of legumes on wheat as influenced by nitrogen fertilization. *Agric. Sci. Res. J.*, **2**(4): 167-176.

Gangwar, B.; Ravisankar N., Vijayabaskaran, S. and Vishwanath, A.P. 2013. On farm nutrient response of crops and cropping systems. *Project Directorate for Farming Systems Research, Modipuram*, Meerut, U.P.

Gaur, A.C., Nilkantan, S. and Dargan, K.S. 2002. Organic Manures, ICAR, New Delhi, India.

Gupta, Meenakshi, Bali Amarjit, S., Kour, Sarabdeep, Bharat, Rajeev and Bazaya, B. R. 2011. Effect of tillage and nutrient management on resource conservation and productivity of wheat (*Triticum aestivum*). *Indian Journal of Agronomy*, **56**: 116-120.

Gupta, R. K. 2003. Is conventional tillage essential for wheat. Pp 95-100. Addressing Resource conservation issues in rice-wheat systems of south Asia-A resource book. Published by Rice wheat Consortium for Indo-Gangetic plains.

Hegde, D. M. 1998. Effect of integrated nutrient management on productivity and soil fertility in pearl millet (*Pennisetum glaucum*)-wheat (*Triticum aestivum*) cropping system. *Indian J. Agron.* **43** : 580-587.

Hegde, D. M. and Katyal, V. 1999. Long term effect of fertilizer use on crop productivity and soil fertility in pearl millet – wheat cropping system in different agro-ecoregions. *J. Maharashtra Agric. Univ.* **24** : 16-20.

Hegde, D.M. and Babu, S.N.S. 2004. Role of balanced fertilization in improving crop yield and quality. *Fertiliser News.* 49(12): 103-110, 113-114, 131.

Hegde, D.M. and Katyal, V. 1998. Long term effect of fertilizer use on crop productivity and soil fertility in maize (*Zea mays*)-wheat (*Triticum aestivum*) cropping system in two soil types. *Indian J. Agric. Sci.* 68(2): 4-6.

Hegde, D.M., Sreenath, P.R., Bhatnagar, K.C., Kaur, R., Kaur, A., Kumar, M. and Sharma, N.M. 1992. Cropping System Research Annual Results. 1991-92, p. 321. Project Directorate for Cropping System Research, Modipuram.

Hobbs, P. and Morris, M. 1996. Meeting South Asia's future food requirements from rice-wheat cropping systems: priority issues facing researchers in the post-green revolution era. Natural Resource Group Working Paper. 96.01. CIMMYT, Mexico.

Hobbs, P. R., Giri, G. S. and Grace. P. 1997. Reduced and zero tillage options for the establishment of wheat after rice in south Asia. Rice wheat Constorium paper Series 2. Rice- Wheat Constorium for the Indo-Gangetic Plains, New Delhi, India.

Jackson, M.L. 1973. *Soil Chemical Analysis.* Prentice Hall of India Pvt. Ltd., New Delhi.

Kanwarkamla. 2000. Legumes- the soil fertility improver. Indian Farming, 50(5): 9.

Katyal, V., Gangwar, B. and Gangwar, K. S. 1999. Long term effect of integrated nutrient supply on yield stability and soil health under pearl millet –wheat cropping system. *J. Maharashtra Agric. Univ.* **24** : 143-146.

Katyal, V., Gangwar, B. and Gangwar, K.S. 2002. Yield trends and soil fertility changes in pearl millet–wheat cropping system under long term integrated nutrient management. *Ann. Agric. Res.* (*New Series*) 23: 201-205.

Khurana, H.S., Singh, Yadvinder. 2008. Site specific nutrient management performance in a rice-rice-wheat cropping system. *Better Crops with Plant Food.* **92**(4): 26-28.

Kukal, S.S.; Rasool, R.; Benbi, D.K. 2009. Soil organic carbon sequestration in relation to organic and inorganic fertilization in relation to in rice-wheat and maize-wheat systems. *Soil Till. Res.* **102**: 87-92.

Kumar, Pawan, Nanwal, R.K. and Yadav, S.K. 2005. Integrated nutrient management in pearl millet (*Pennisetum glaucum*)-wheat (*Triticum aestivum*) cropping system. *Indian J. Agric. Sci.* 75(10): 640-43.

Laxminaryana, K. and Patiram. 2006. Effect of integrated use of inorganic, biological and organic manures on rice productivity and soil fertility in Ultisols of Mizoram. *J Indian Soc. Soil Sci.* **54**: 213-220.

Little, T. M. and Hills, F. J. 1978. Agricultural Experimentation.New York. NY : John Willy and Sons.

Majuakim, L. and Kitayama, K. 2013. Influence of polyphenols on soil nitrogen mineralization through the formation of bound protein in tropical montane forests of Mount Kinabalu, Borneo. *Soil Bio. Biochem.* **57**:14-21.

Olsen, S.R., Cole, V.C., Watanabe, F.S. and Dean, L.A. 1954. Estimation of available phosphorus in soils by extraction with sodium bicarbonate. *Cir. U.S. Dep. Agric.* 939.

Oo Lar Mar Naw; Shiva, Y.S. and Kumar, Dinesh. 2007. Effect of nitrogen and sulphur fertilization on yield attributes, productivity and nutrient uptake of aromatic rice (*Oryza sativa*) *Indian J. Agric. Sci.* **77**: 772-775.

Panse, V.G. and Sukhatme, P.V. 1985. *Statistical Methods for Agricultural Research Workers*. IVth edition, ICAR, New Delhi.

Patel, D.M., Patel, S.K. and Karelia, G.N. 1995. Integrated nutrient management in pearl millet (*Pennisetum glaucum*)-wheat (*Triticum aestivum*) cropping system. *Indian J. of Agron.* 40: 266-268.

Prasad, R. 1998. A Practical Manual for Soil Fertility. New Delhi : Division of Agronomy, Indian Agricultural Research Institute.

Ruhal, D.S. and Shukla, U.C. 1979. Effect of continuous application of farmyard manure and nitrogen on organic carbon and available N, P and K content in soil. *Indian J. Agric. Chem.* **12**: 11-18.

Saroch, K., Bhargava, M. and Sharma, J.J. 2005. Diversification of existying rice (*Oryza sativa*) – based cropping systems for sustainable productivity under irrigated conditions. Indian J. of Agron. 50(2): 86-88.

Singh, G., Singh, V.P., Singh, O.P. and Singh, R.K. 1997. Production potential of various cropping systems in looedprone areas of eastern Uttar Pradesh. Indian J. Agron. 42(1): 9-12.

Singh, Raj, Singh, Bhagwan and Patidar, M. 2008. Effect of preceding crops and nutrient management on productivity of wheat (*Triticum aestivum*)-based cropping system in arid region. *Indian J. Agron.* **53**(4): 267-272.

Singh, S. S., Prasad, L. K. and Upadhyaya, A. 2006. Root growth, yield and economics of wheat (*Triticum aestivum*) as affected by irrigation and tillage practices in south Bihar. *Indian Journal of Agronomy*, **51**: 131-134.

Subbiah, B.V. and Asija, G.L. (1956). A rapid procedure for estimation of available nitrogen in soils. *Curr. Sci.* 25: 259-260.

Verma, L.P.; Yadav, D.S. and Singh, Room. 1987. Effect of continuous cropping and fertilizer application on fertility status of soil. *J. Indian Soc. Soil Sci.* **35**: 756-766.

www.ingramcontent.com/pod-product-compliance
Ingram Content Group UK Ltd.
Pitfield, Milton Keynes, MK11 3LW, UK
UKHW021948270726
14060UKWH00002B/422